Springer Theses

Recognizing Outstanding Ph.D. Research

For further volumes:
http://www.springer.com/series/8790

Aims and Scope

The series "Springer Theses" brings together a selection of the very best Ph.D. theses from around the world and across the physical sciences. Nominated and endorsed by two recognized specialists, each published volume has been selected for its scientific excellence and the high impact of its contents for the pertinent field of research. For greater accessibility to non-specialists, the published versions include an extended introduction, as well as a foreword by the student's supervisor explaining the special relevance of the work for the field. As a whole, the series will provide a valuable resource both for newcomers to the research fields described, and for other scientists seeking detailed background information on special questions. Finally, it provides an accredited documentation of the valuable contributions made by today's younger generation of scientists.

Theses are accepted into the series by invited nomination only and must fulfill all of the following criteria

- They must be written in good English.
- The topic should fall within the confines of Chemistry, Physics, Earth Sciences, Engineering and related interdisciplinary fields such as Materials, Nanoscience, Chemical Engineering, Complex Systems and Biophysics.
- The work reported in the thesis must represent a significant scientific advance.
- If the thesis includes previously published material, permission to reproduce this must be gained from the respective copyright holder.
- They must have been examined and passed during the 12 months prior to nomination.
- Each thesis should include a foreword by the supervisor outlining the significance of its content.
- The theses should have a clearly defined structure including an introduction accessible to scientists not expert in that particular field.

At Venus Ascanio placidam per membra quietem inrigat, et fotum gremio dea tollit in altos Idaliae lucos, ubi mollis amaracus illum floribus et dulci aspirans complectitur umbra

Eneide, Libro I, 691-694

Supervisor's Foreword

The increasing complexity and interconnectedness of world economies and financial markets call for a radical paradigm change in the actual economic thinking. The 2007 financial crisis, the following economic stagnancy, and the very slow economic recovery of western countries have shown all the weaknesses of standard economic theories and standard economic policies.

The progress of economic disciplines has now reached a turning point and new non-orthodox paths are emerging in opposition to mainstream approaches. A key point of this new vision is that it cannot be anymore neglected that economies are complex systems and, given the present level of interconnectedness, economies and markets must be described at a global level.

Any new approach aiming at an effective description and comprehension of economic systems must include these features. In this perspective, physicists and in general interdisciplinary approaches can contribute to this paradigm change in a twofold way.

On one hand, economics must really incorporate the fact that economy is a complex system in an effective way and conceptual frameworks from Complex Systems Theory in Physics can give new insights into complexity-oriented descriptions of economic systems.

On the other hand, Economics must be turned into a much more experimental and data-oriented discipline. Data-driven methods and tools must be developed to properly address the nowadays economic complexity and concepts like international externalities. The several economic theories introduced over the years are now considered as global paradigms which compete against each other, aiming at being universal recipes for all situations and regimes of present economic systems. Economics must instead point to a scenario in which the various theories have different degrees of effectiveness depending on the economic regime under consideration as it happens in Physics.

The ambitious goal of this thesis was to introduce new quantitative measures (i.e. metrics) to turn economic system description from qualitative to quantitative. Financial markets are a natural starting point for a scientific approach to this discipline since the large amount and availability of high-frequency records of stock exchanges. On this account, agent-based models have shown the weakness of standard theories of financial markets in terms of representative and homogeneous agents and rational expectation and qualitatively shown the importance of concepts such as heterogeneity, herding, and nonlinear feedbacks in order to address the complex dynamics and features of financial markets. However, these kinds of models are usually as black boxes where it is almost impossible to answer the question *what gives what*. Matthieu has participated in the development of a minimal and workable agent-based model which demonstrated that the statistical features of markets (i.e. Stylized Facts) are a finite size effects in this class of models and correspond to the regime toward which markets self-organize. Self-organization of financial markets is often neglected by the literature in this field even if it is a key point to understand the stability of markets. Agent-based models, however, represent only a conceptual and qualitative metaphor which cannot, at the present stage, be turned into concrete tools or metrics for economic regulatory policy, systemic risk estimates, etc. In this thesis a possible path in order to make agent-based models effective tools for the description of financial markets is indicated and discussed in terms of effective number of agents.

The second issue addressed in this thesis is the challenge of measuring the competitiveness of countries and the complexity of products starting from an export flow dataset. Although economics is far from being a good candidate for experimental approach given a systematic lack of reliable data, a recent strand of data-driven literature has proposed a complexity approach to international trade highlighting the fact that relatively rich countries are also characterized by high diversification of their export basket in contrast to standard Ricardian arguments in terms of competitive advantage leading to specialization of productive systems. The challenge is therefore assessing quantitatively advantage of diversification for the potential of development and growth of countries. Matthieu has been one of the main actors in the development of a novel statistical approach based on coupled nonlinear maps, whose fixed point defines new non-income based metrics for the country fitness and product complexity. This work appears to be groundbreaking in many directions and aspects. As an example, through this new competitiveness metrics it is possible to compare non-income based factors of fitness and complexity to income measures, for instance, countries' GDPs *per capita*. Preliminary results reveal that the difference between the two sources of information is crucial for quantitatively assessing the hidden potential of growth of countries. In addition, the analysis of the dynamics of countries' trajectories in the Fitness-Income plane shows the existence of several regimes of economic growth calling for a description in terms of a heterogeneous dynamics of the economic complexity.

In summary, Matthieu carried out frontier and, in my opinion, groundbreaking researches in the perspective of the paradigm change needed in the present economic thinking. The development of quantitative tools and metrics of Matthieu's research activity points exactly in this direction and represents a fundamental contribution to economic theory.

Rome, April 2013

Prof. Luciano Pietronero

Acknowledgments

I would like to thank Prof. Pietronero, my tutor during my Master's Degree and my Ph.D. activity, for the countless conversations about Science during which I have learnt the enthusiasm for the search for Nature's answer.

I thank all the members of the PIL group for all the suggestions received.

I thank all my colleagues (and, above all, friends before being colleagues) of the Ph.D. program delocalized among the New and Old Building of the Physics Department.

Among these friends, there is a special Amica (i.e. special friend in italian…) who I would like to thank for about 2900+ days of ordinary madness.

All the people who stayed in Room 206 and made my Ph.D. unforgettable deserve special thanks.

The Room 206 also deserves nostalgic thanks: this room is full of personal memories, some unforgettable, some pleasant , some others frustrating and sad, but all equally important.

I acknowledge the Italian PNR project CRISIS-Lab and FET Open project nr. 255987 FOC (Forecasting Financial Crisis) for the support given to my work.

Last but not least, I would like to thank my family which supported and put up me (in Italian this phrase sounds like a wordplay since supported is translated by supportare while put up by sopportare).

Abstract

The fundamental idea developed throughout this work is the introduction of new metrics in Social Sciences (economics, finance, opinion dynamics, etc.). The concept of metrics, that is the concept of measure, is usually neglected by mainstream theories of economics and finance. Financial markets are the natural starting point of such an approach to Social Sciences because a systematic approach can be undertaken and the methods of Physics have shown to be very effective. In fact since a decade there exists a huge amount of high frequency data from stock exchanges which permit to perform *experimental procedures* as in Natural Sciences. Financial markets appear as a perfect playground where models can be tested and where repeatability of empirical evidences are well-established features differently from, for instance, Macro-Economy and Micro-Economy. Thus, Finance has been the first point of contact for the interdisciplinary application of methods and tools deriving from Physics and it has also been the starting point of this work.

We investigated the origin of the so-called Stylized Facts of financial markets (i.e. the statistical properties of financial time series) in the framework of agent-based models. We found that Stylized Facts can be interpreted as a finite size effect in terms of the number of effectively independent agents (i.e. strategy) which results to be a key variable to understand the self-organization of financial markets.

As a second issue we focused our attention on the order book dynamics both from a theoretical and a data-oriented point of view. We developed a zero intelligence model in order to investigate the role of vanishing liquidity in the price response to incoming orders. Within the framework of this model we have analyzed the effect of the introduction of strategies pointing out that simple strategic behaviors can explain bursts of intermittency and long memory effects. On the other hand we quantitatively showed that there exists a feedback effect in markets called *self-fulfilling prophecy* which is the mechanism through which technical trading can exist and work. This feature is a very interesting quantitative evidence of a self-reinforcement of agents' belief. Last but not least, nowadays we live in a computerized and networked society where many of our actions leave a digital trace and affect other people's actions. This has led to the emergence of a new data-driven research field. In this work we highlight how non-financial data can be used to track financial activity, in detail we investigate query log volumes, i.e., the volumes of searches for a specific query done by users in a search engine, as a proxy for trading volumes and we find that users' activity on Yahoo! search engine anticipates trading volume by one to two days.

Differently from Finance, Economics is far from being an ideal candidate to export the methodology of Natural Sciences because of the lack of empirical data since controlled (and repeatable) experiments are totally artificial while real experiments are almost incontrollable and non-repeatable due to a high degree of non-stationarity of economic systems. However, the application of method deriving from complexity to the economic theory of growth is one of the more important achievements of the work developed here. The basic idea is to study the network defined by international trade flows and introduce a (non-monetary) metrics to measure the complexity and competitiveness of countries' productive systems. In addition, we are able to define a metrics for products' quality which overcomes traditional economic measures for the quality of products given in terms of hours of qualified labor needed to produce a good. The method developed provides some impressive results in predicting economic growth of countries and offers many opportunities for improvements and generalizations.

Contents

Chapter 1
Introduction

Why did physicists start to be interested in economics, finance, and in general in the social sciences? Why and how can physics make an effective contribution to the description and understanding of these disciplines?

The answer can be traced back to the differences between physics and economics (and to some extent the social sciences). Physics typically proceeds through a virtuous circle usually called the scientific method: theoretical and experimental approaches are parallel, complementary, they progress by an ongoing reciprocal feedback and they are really *two sides of the same coin*. Therefore physics is both an experimental and theoretical science.

On the other hand, standard approaches in the social sciences and economics traditionally neglect the experimental side of the coin. Indeed, the social sciences belong to those sciences defined as observational. Observational sciences are disciplines that operate in a field where controlled observations cannot be used in order to study causes and effects. Unavoidably, social phenomena are usually characterized by a high degree of non-stationarity and a low level of controllability and reproducibility of experiments.

However, we argue that in those disciplines there is a systematic lack of data-oriented approaches and especially of *metrics* in order to give a quantitative description of social phenomena. By the absence of metrics, we mean that in The social sciences, and especially in economics, the introduction of the concept of *distance* to make a phenomenon quantitative is often neglected or does not appear as a crucial point. For instance, quantities such as the systemic risk of financial markets, the mood of people, the quality/complexity of a product, the competitiveness of a country, etc., can hardly be measured by the tools provided by mainstream economic theories.

From this observation, we argue that the economic mainstream cannot be defined as an experimental science and not even an observational one, given the actual state of the art. We are aware that there exist some publications in economics where the interplay between empirical evidence and models is not marginal but they are usually neglected or never incorporated in mainstream economic theories.

M. Cristelli, *Complexity in Financial Markets*,
Springer Theses, DOI: 10.1007/978-3-319-00723-6_1,

It becomes apparent, from these considerations, what contribution physics can make to economics. Physics can contribute to making economics an observational and experimental science. In particular, methods and tools derived from physics and mathematics can be effective in providing, and turning from qualitative to quantitative, the description of social and economic phenomena. Thus by introducing effective metrics for economic systems, twofold progress would be achieved: on one hand, data-oriented approaches and economic theories would have a common language to achieve mutual feedback and, on the other, key issues for policy modeling and policy making could be properly addressed and suitable interventions could be undertaken.

The absence of mutual feedback between data and theories in economics has led this discipline towards two problematic points with respect to the *desirable* forecast skills that should be required of this field given its societal impact:

1. On one hand economics has basically developed only mathematically solvable models, which are typically very far from the scenarios depicted by empirical analysis. If a similar approach were adopted in physics, only the hydrogen atom would exist!
2. On the other hand the very poor feedback between theoretical models and data-driven studies has led to inadequate or partial measurement tools (i.e., mathematical tools) for economics and finance. In fact these tools are usually defined only on the basis of the model's results. The attempts to give a quantitative description of economic systems, to some extent, reflect only the theoretical *a priori* interpretation of these and never the empirical evidence. Tools are modeled to be optimal proxies only for what is expected from models and not, as happens in physics, for what is really observed. In summary, very often in economics, unlike in physics, the only driving forces for the discipline are mainstream theories and the problem of empirical validation is completely ignored. For instance, in recent work [1–3], the standard forecast framework of linear regressions appears to be the wrong tool to quantify the forecast of the growth and development of country economies.

We want to point out that we are not rejecting all economic theories; our criticisms are mainly aimed at the methodology of economics with respect to empirical data. Economics needs a revolution [4] but this does not imply that economic theories are all and completely wrong. Physicists (and in general non-economists) are not tempting to discard all current economic theories, on the contrary, we believe that some of the current theories can be seen as the zero order for new contributions that can be made to the social sciences—especially finance and economics—by physics. The current *mean field* theories of economics cannot be the arrival point of this discipline but the starting one for the revolution that is now necessary given the powerlessness of traditional previsional tools in the face of the last decade of financial crises and the following economic stagnancy.

Let us give a simple example to provide insight into our considerations. In the 1970s Black, Scholes and Merton [5] introduced their theory to estimate the price of a derivative given the price of the underlying stock. Black and Scholes's equation has been the zero-order theory for thirty years of financial engineering, which has developed very complicated stochastic processes and models to try to always assess in a better way the problem of derivative pricing. However, Black and Scholes's equation represents only the starting point and the first and somehow *naive* attempt to price derivatives. In fact the financial crisis that occurred some years after the introduction of this equation showed all the limits of this modeling in terms of just a simple multiplicative random walk. Financial crises are a special kind of empirical feedback since if the models are not predictive, people lose money. Consequently, it is not surprising that financial engineering is the most data-oriented discipline of economics, even if the problem set by systemic risk of financial markets is far from being solved since nowadays we do not have real and reliable tools to estimate this kind of risk.

The Black and Scholes theory is also a perfect example of how an interdisciplinary approach can be fruitful. In fact this equation represents a real breakthrough in the field of financial instruments but the equation itself was not a *new* discovery or a *new* theory; the real novelty was its application. In fact, mathematicians and physicists have known about multiplicative stochastic processes since the first two decades of the 20th century, in physics processes of this kind are usually known in the Langevin formalism, while in mathematics multiplicative processes arose when stochastic calculus had been formalized.

In the last 20–30 years the basis of a paradigm shift in economics and in general in the social sciences has arisen: new sources of data emerged and now permit, at least in some fields of the social sciences and economics, the lack of empirical data to be overcome. So-called big data science and ICT (information and communications technology) have disclosed such a huge amount of raw data about social and economic phenomena that now the invoked revolution of the methodology of economics is possible.

This also explains why financial markets appeared as the first candidate for this interdisciplinary application of physics: they allow a systematic approach to be taken. In fact, for two decades there has existed a huge amount of high frequency data from stock exchanges, which permits *experimental procedures* to be performed as in the natural sciences. Financial markets appear to be a perfect playground where models can be tested and where repeatability of empirical evidence is a well-established feature. In addition the methods of physics have been very effective in this field and have also often given rise to concrete (and sometimes profitable) financial applications.

Nevertheless, until now, economic complexity has not led to a concrete change in the economic mainstream. We believe this is due to the fact that the revolution invoked in [4] has taken place mainly at a qualitative and conceptual level. As an example, agent-based models have been a fundamental turning point for economic theories of financial markets. They allowed the market description in terms of homogeneous agents, rational expectation, cause–effect relation in price evolution and price equilibrium to be surpassed and revealed the importance of concepts such as belief and strategy heterogeneity, herding and non-rational behavior to explain the complex features of the dynamics of financial markets.

This new perspective unfortunately takes place only at a metaphoric level and the agent-based models, even if they are able to explain the stylized facts from a qualitative point of view, do not provide concrete tools to address key issues of markets such as risk.

In the remainder of this overview we discuss prospects for the development of economic complexity into a mature discipline. In particular, after a short summary of the state-of-the-art of agent-based modeling, we depict some possible paths by which this class of models could be turned into concrete tools for the measurement of market risks.

We also give some insights into new perspectives of economic complexity beyond financial markets. The availability of new datasets (so-called big data science) has recently opened up new domains—traditionally characterized by an intrinsic lack of empirical data—to which methodology from physics and complex systems can be fruitfully applied. This can be achieved through the introduction of metrics and quantitive methods in economics, and as an example of this frontier of research in economic complexity we review how archival data on export flows can be turned into a concrete tool to assess the competitiveness of countries and the complexity of products.

1.1 The Revolution of Economics Needs a Revolution

The major contributions of economic complexity to financial markets are twofold: phenomenological and effective description of the dynamics of markets at a micro-level and conceptual comprehension of stylized facts.

The former contribution provides fundamental insights into the non-trivial nature of the stochastic process performed by stock prices [6–10] and into the role of the interplay between agents to explain the behavior of the order impact on prices [10–17].

The latter approach has instead tried to improve upon the traditional economic models based on concepts such as price equilibrium and homogeneity of agents in order to investigate the role of heterogeneity of agents/strategy with respect to the price dynamics [18–25].

In the next sections we are going to focus on this second contribution to finance and we discuss how to turn this approach into quantitive tools for key issues of financial markets.

1.1.1 Agent-Based Modeling: The State-of-the-Art

The standard theory of economics for financial markets is based on the following elements:

- Situation of equilibrium with equal (representative) agents that are (quasi)rational, have the same information, and process it in the same way.

- Important price changes correspond to new information that arrives on the market. The fact that this information is random and independent leads to the famous random walk model and the corresponding Black and Scholes equations.
- This new information modifies the ratio between offer and demand and then also the price. This corresponds to a mechanical equilibrium of the market.
- These concepts also imply a cause–effect relation, in which large price changes are due to the market reaction to the arrival of exogenous important news. Therefore a large price change is supposed to be associated with an equally large exogenous event.

It is too easy to argue that most of these assumptions have no basis at all. So one may wonder why they are so widely adopted. The real reason is that they permit an analytical treatment of the problem. This is a very different perspective from natural sciences, in which very few realistic problems can be treated exactly. In addition these ideas are usually not tested against empirical data, a fact that strongly limits the scientific basis of the whole framework. The most reasonable of the above assumptions is the fact that indeed many external news items are random and incoherent so the random walk appears reasonable as a simple model. On the other hand the assumptions made for the behavior of the agents are very far from reality. The agents can be very different from each other, their level of information and the way the use it are also very different. Finally they are not at all independent and, especially in situations of crisis, the rationality hypothesis can be seriously wrong. In these situations fear and panic, as well as euphoria, lead to very strong herding behavior, which is completely neglected in the standard model. This situation naturally calls for a possible description in terms of critical phenomena and complex systems. The study of complex systems refers to the emergence of collective properties in systems with a large number of parts with interactions between them. This implies a change of paradigm from the previous mechanical model to a complex model in which intrinsic instabilities can develop in a self-organized way without a cause–effect relation.

This change of paradigm is witnessed by the fact that in past years there has been great interest in the development of agent-based models aimed at reproducing and understanding the stylized facts observed in the financial time series. Agent-based models represent a broad class of models that have been introduced to describe the economic dynamics in a more realistic way. Their building blocks are:

- the agents are heterogeneous with respect to their various properties such as strategies, wealth, and time scale;
- the interaction between them is a fundamental element and, of course, it can have many different characteristics;
- price dynamics depends on the balance between offer and demand but the specific implementation can be different.

Usually the dynamics of models of this type is able to produce deviations from the Gaussian behavior related to the stylized facts for some specific range of parameters. The main criticisms of agent-based models rely on the fact that up to now these models have not been able to produce concrete tools for measuring risk. They are somehow conceptual metaphors that showed that the mainstream of economic theories on financial markets were inadequate. They certainly revolutionized economics but a new step is now required to make agent-based models a concrete tool to properly address issues such as systemic risk and market instabilities. The path towards this new step is one of the new frontiers of economic complexity.

1.1.2 Agent-Based Modeling: Perspectives

There exists a point that is usually neglected by agent-based models: The stylized facts usually correspond to a particular situation of the market. If the market is pushed outside such a situation, it will evolve to restore it spontaneously. The question is why and how all markets seem to self-organize in this special state and, even more importantly, when do markets become less stable and can be more likely pushed far from this special state? The answer to this question conceptually represents a fundamental point in understanding the origin of the stylized facts and from a practical point of view the starting point for the development of a quantitative measure for systemic risk.

In this respect, in a series of recent theoretical publications [18–21, 26], we show that stylized facts correspond to a finite size effect in the sense that they disappear in the limit of large N (total number of agents) or large time. Given the general nature of these results we may conjecture that these should apply to a broad class of agent-based models and they are not a special property of a specific model. This situation has important implications both for the microscopic understanding of the stylized facts and for the analysis and interpretation of experimental data. In this series of papers it is shown that a price that is very stable corresponds to a very large number of market participants. On the other hand a small number of agents leads to large movements in the price. In other words, when markets are in a low-volatility state agents tend to operate incoherently and the impact of an agent's action is negligible. Instead when the number of independent agents/strategies diminishes, the market is evolving towards a situation in which almost all the agents react in the same way. In such a scenario, on one hand the market susceptibility increases enormously, giving rise to potential systemic instabilities. On the other hand the more the agents act coherently, the more the amplitude of the market fluctuations grow, triggering a self-reinforcing effect. From an empirical point of view a number of studies confirm that markets oscillate between states characterized alternatively by mean reverting (stable) and mean escaping tendencies (large fluctuations, market instabilities). That being so, the empirical comprehension of herding behaviors and the dynamics of the number of effective agents represent the second step to completing the paradigm change introduced by agent-based models and to turning them into concrete predictive tools to help policy making to prevent global financial crashes.

The background of such a vision is twofold:

- We believe that this dynamics (number of independent strategies) may anticipate the market instabilities, and a measure of the number of effectively independent strategies can be a concrete early warning that can be used to detect them. We want to stress that we do not aim to predict when the next systemic crash will occur but when potential conditions for large instabilities arise.
- As discussed previously, a weakness of agent-based models is that the experimental framework defined by the stylized facts is rather limited and an improvement of this body of knowledge appears to be the bottleneck in the field. With the knowledge of the time series of the number of independent agents, these models can be tuned, for the first time, using a quantity that has a direct and clear counterpart in the model: the number of agents. In such a way, agent-based models can be turned into powerful instruments to give a complete scenario in terms of their ingredients: herding parameters, trust, kind of strategies, liquidity, and in general all the elements inserted in these models.

1.1.3 Microstructure of Financial Markets

The availability of the high-frequency records of stock exchanges has led to a better description and understanding of price dynamics, especially for price formation, price response, and order impacts. This huge amount of data has also improved so-called financial engineering techniques. This data also provided the basis for the development of very effective and powerful phenomenological descriptions of market activities, which give some insight into market risk. However, systemic risk of financial systems is far from being a solved problem. A complementary approach to the analysis of microstructure to address system risk is represented by network-based techniques [27], which looks at the robustness/fragility of credit and financial networks.

1.1.4 Non-financial Proxies

An interesting data-driven approach to the modeling of economic and financial systems is represented by the recent availability of large datasets, which ranges from the web activity of search engine and social networks to human activity records such as phone, car usage, and international trade flows. So-called global systems science sees the systems in which we live as a series of interacting networks. This networked reality has favored the emergence of a new data-driven research field where mathematical methods of computer science, statistical physics, and sociometry provide quantitative insights into a wide range of disciplines [1, 28–36]. In this perspective, big data may represent an important source of informative content to improve the comprehension and description of economic systems and social phenomena. The last 5–10 years have delivered a series of pioneering publications on this subject, which showed the potential of these non-standard sources of data. We briefly list below the main results in this very young field and in the next section we discuss in detail a recent piece of work.

- User web activity can be turned into a useful tool to track the spreading and evolution of flu and other epidemics.
- Web search volumes can be predictive on trading volumes.
- Web searches provide informative content about customer habits and trends and can permit the tracking of sales in various sectors, such as phone activations, car sales, and tourism.
- Human mobility and human dynamics in highly populated areas such as metropolises can give key hints that can be projected to so-called smart cities.
- Import–export archival datasets collected by customs represent an important source of data to quantify the economic competitiveness of countries and the complexity of products.

1.2 Economic Complexity: From Finance to Economics

Unlike finance, economics is far from being an ideal candidate to which to export the methodology of natural sciences because of the lack of data, since controlled (and repeatable) experiments are totally artificial while real experiments are almost uncontrollable and non-repeatable owing to the high degree of non-stationarity of economic systems. However, the application of methods deriving from complexity theory to the economic theory of growth is one of the most important achievements of a stream of publications that have appeared in the last five years [1, 2, 34, 35].

A concrete example of how data-driven approaches can lead to new methods and tools for economics is the network defined by international trade flows [35]. It is indeed possible to introduce [1] a (non-monetary) metric to measure the complexity and competitiveness of the productive system of countries. In addition it is possible to define a metric for product complexity that surpasses the traditional economic measure for the quality of products given in terms of hours of qualified labor needed to produce a good. The method developed provides some impressive results in predicting economic growth of countries and offers many opportunities for improvements and generalizations [1, 3, 36].

1.2.1 Country Competitiveness and Product Complexity

Which is the country with the highest competitiveness in the ecosystem defined by the world trade web? Which one has the most complex productive system?

Some recent publications [34, 35] have indicated new approaches to macroeconomic issues, inspired by network-based methods. The basic idea is to study the networks defined by international trade flows [37] and to gain information about the productive structure of countries and its future development. In this kind of analysis, methods derived from physics have proven to be extremely effective.

According to mainstream theories on economic growth [38–42], of which Adam Smith is the father, specialization should provide an evolutionary advantage for a country. In fact, citing Adam Smith "it is the great multiplication of the productions [...] in consequence of the division of labour, which occasions [...] that universal opulence which extends itself to the lowest ranks of the people." This claim can easily be tested by looking at the binary country–product (C-P) matrix. In the C-P matrix each entry is 1 if a particular country C is a significative exporter of product P, 0 if it is not. Specialization would mean that when the rows and columns of the C-P matrix are arranged properly, it should appear as a block-diagonal matrix. Observations disagree with this prediction and the C-P matrix is found to be approximately triangular.

It is worth noticing that for biological organisms and species there is evidence that diversification usually gives an evolutionary advantage compared to specialization, in particular, too-specialized species are weaker when global changes occur and tend to become extinct. Therefore it appears that in a highly interconnected and dynamic environment, as modern globalized economies are, country competitiveness resembles the biological fitness of species.

The existence of significative off-diagonal terms in the C-P matrix implies that specialization does not appear as a natural strategy for countries. Furthermore, the triangular shape gives an additional and specific piece of information about the correlation between the composition of the export basket and the kind of products exported by a country. In fact we find that some countries have a large diversification of their production and consequently make almost all products, while scrolling down the rows of the matrix, the number of exported products decreases and countries become more and more specialized on a small subset of products that are exported by almost all countries. Furthermore the most diversified countries coincide with the richest ones, while the poorest ones are the ones specialized on those few products exported by almost every country. These empirical findings suggest that diversification is more important than specialization for country growth, as for biological systems. This is not surprising since specialization may be preferred or provide advantages only in a static and in-equilibrium system. On the other hand in an out-of-equilibrium and dynamic playground where different productive systems compete, diversification becomes an evolutionary advantage in the case of changes. The fact that the richest countries are also the most diversified ones confirms this hypothesis.

The question that now arises is how to measure the complexity of a productive system given the export basket. A recent paper [35] proposes a new conceptual framework in order to explain why diversification is an optimal strategy and why the C-P matrix is basically triangular. The authors hypothesize that there exist some special endowments of the countries, called capabilities, which are non-exportable and stay localized in the nation where they are. The authors of [35] assume that these capabilities are the key element in grasping the complexity of a productive system because each product requires a specific set of capabilities which must be owned by a country in order to produce and then export it. The first step of our work is then to define a consistent and solid way to infer the level of capabilities that each country has. We call this measure "fitness". For products we define a corresponding measure, the "complexity".

Fitness and Complexity The theory of capabilities suggests a strategy for using the information contained in the C-P matrix in order to determine fitness and complexity in a self-consistent, recursive way. At order 0 a natural guess for the fitness of a given country is to consider its diversification: the more products it exports, the more capabilities it should have. This is easily generalized when the diversification is weighted with the complexity of products in a simple linear way.

Concerning products, the measure has to be different because a strong nonlinearity is needed to take into account the following: considering a product of two very different countries (with respect to fitness), say Germany and Nigeria, its complexity cannot simply be obtained by averaging the level of fitness of the two countries, because the signal that Nigeria is able to produce it means that its capabilities (which are very few) are enough to enable the production of that product. Thus in the evaluation of the complexity a strong nonlinearity must be involved, in order to make the "worst" producer of a given product the most important in the determination of complexity. This has to be achieved by still considering that an ubiquitous product is likely to be of low complexity. These two characteristics have to be consistently mixed. Given these qualitative definitions, each one involving the other, a recursive procedure (with a proper renormalization) leads to an asymptotic, stable distribution for fitness and complexity.

Fundamental Analysis of Countries and Complexity The measure developed in [1] allows a new kind of fundamental analysis of the growth potential of countries to be defined. First, combining information about the complexity of products with the explicit normalized volumes of export of each single country, it is possible to realize a "spectroscopy" of each productive system, and, using information about the dynamics of such spectra, infer their future structure. Moreover, it is possible to combine the information about the fitness of a single country and its competitors for any single product, and consider that product's complexity in order to define measures of "competitiveness".

Through this new algorithm and the associated metrics it is possible to compare non-monetary factors of fitness and complexity to measures of economic intensity, for instance the countries' GDPs per capita. We argue that countries that show both a high fitness and a high complexity, but a low GDP, are very likely to strongly boost their income in the next decades. This difference reveals crucial information. When the model was retrofitted with data from 1995 to 2010, it was possible to see that the results reflected well what occurred in the real world over that period.

This type of analysis can lead to a novel perspective for the study and industrial planning of countries; it also provides information on the GDP growth and an estimate of the global risk in relation to fundamental economics. It is just the beginning of a novel era for economics in which the large amount of high quality data and some original ideas open the way to more consistent methods and permit us to move from qualitative to quantitative considerations, shedding new light on important and unexpected properties.

1.3 General Frame of this Thesis

The fundamental idea developed throughout this thesis on economic complexity is that a revolution of the revolution of economics is needed to turn this field into a mature discipline. The first revolution of economic complexity [4] led to a conceptual paradigm shift, and agent-based models have shown, from a qualitative point of view, the crucial role played by concepts such as agent heterogeneity and herding behavior to understand the non-trivial features of financial time series. The second revolution must lead the paradigm shift from a conceptual and qualitative level to a quantitive and effective description of economic systems. This can be achieved through the introduction of new metrics and quantitive methods in the social sciences (economics, finance, opinion dynamics, etc.). In fact, the concept of metrics is usually neglected by mainstream theories of the economy and finance. Only in that way can economic complexity concretely affect the thinking of the economics mainstream and, in this sense, become a mature discipline. The large availability of datasets (so-called big data science) has recently revealed a new, promising path towards such perspectives and, as an example, we have briefly discussed how archival data about export flows can be turned into a concrete tool to assess the competitiveness of countries and the complexity of products.

All in all, the fundamental idea underlying all the work presented in this thesis is the introduction of novel metrics for the social sciences and specifically for economics and finance. We have said that the natural starting point of such an approach to the social sciences are financial markets, where a systematic approach can be undertaken and the methods of physics are very effective. Thus it was also the starting point of this work.

We investigated the origin of the so-called stylized facts of financial markets (i.e., the statistical properties of financial time series) in the framework of agent-based models. We found that stylized facts can be interpreted as a finite size effect in terms of the number of effectively independent agents, which turns out to be a key variable for understanding the self-organization of financial markets [18–21, 26].

As a second issue we focused our attention on the order-book dynamics from both a theoretical and a data-oriented point of view. We developed a zero intelligence model [43–45] in order to investigate the role of vanishing liquidity for the price response to incoming orders. Liquidity and liquidity fluctuations are key features in order to properly address the systemic risk issue [46].

On the other hand we quantitatively showed that there exists a feedback effect in markets called *self-fulfilling prophecy*, which explains why technical trading can work [47]. This feature is very interesting quantitative evidence for a self-reinforcement of agents' belief.

Finally, nowadays we live in a computerized and networked society where many of our actions leave a digital trace and affect other people's actions. This has led to the emergence of a new data-driven research field [28, 29] that can provide a fundamental contribution to finding and defining suitable metrics for collective human behavior, among which we also list financial market activity [30–32]. In this work [33] we highlighted how non-financial data can be used to track financial activities. In detail, we found that query log volumes, i.e., the volumes of searches for a specific query performed by users in a search engine, can be used as a proxy for trading volume and we also found that users' activity on the Yahoo! search engine anticipated trading volume by one-to-two days.

As stated before, unlike finance, economics is far from being an ideal candidate to which to export the methodology of natural sciences because of the lack of data, since controlled (and repeatable) experiments are totally artificial while real experiments are almost uncontrollable and non-repeatable owing to the high degree of non-stationarity of economic systems. However, the application of methods deriving from complexity theory to the economic theory of growth is one the most important achievements of the work developed here. The basic idea is to study the network defined by international trade flows and introduce a (non-monetary) metric to measure the complexity and the competitiveness of the productive system of countries [35]. In addition we are able to define a metric for products' quality that surpasses the traditional economic measure for the quality of products given in terms of hours of qualified labour needed to produce a good. The method developed provides some impressive results in predicting economic growth of countries and offers many opportunities for improvements and generalizations [1, 36].

1.4 Instructions for the Reader

In this section we describe the organization of this thesis, which is divided into three main parts. In the first part we concentrate our attention on agent-based models. In the second part we focus on data-driven results concerning mainly the dynamics of financial markets and we also present a zero-intelligence model of the order book. In the third part we shift our attention to economics. In detail we investigate and introduce a new metric to *measure* the growth potential of countries' economies.

Finally, in the appendix we highlight an interesting and non-trivial property of Zipfian sets, which must be characterized by a sort of internal consistency/coherence to make Zipf's law valid.

- **Chapter** 2: In this chapter we review the main empirical evidence and regularities of financial markets, which are usually known as stylized facts. These stylized facts reveal that the process followed by stock price cannot be described by a simple random walk or a simple diffusive process since the distribution of returns (i.e., increments) is fat tailed (in detail, it is well described by a power law with exponent between 2 and 4) and the increments are uncorrelated but not independent, in fact the autocorrelation function of the squared increments decay as a power law.

Part I

- **Chapter** 3: We give a detailed overview of the main agent-based models introduced in the last 20 years. This analysis is performed in a schematic way according to seven categories which are the aim of the model, how the agents and their strategies are modeled, the number of the agents and if it can change with time, the evolution rule of the price, which ingredients of the model give the stylized facts (when they are reproduced), the realism versus the tractability of the model and whether, and if so how, the problem of the self-organization of the markets has been taken into account.
- **Chapter** 4: We present a new minimal agent-based model. The main result achieved by this work is a clear answer to the question *what gives what* with respect to the origin of the stylized facts. In fact one of the principal shortcomings from which almost all agent-based models suffer is that they are very complicated *black boxes* that reproduce stylized facts but do not provide a real interpretation of the mechanism underlying the stylized facts. The model discussed in this chapter instead reveals that the stylized facts are a finite size effect in this framework. This observation also permits us to discuss a simple mechanism to explain why markets are self-organized systems in the state of the stylized facts.

Part II

- **Chapter** 5: In this chapter we describe how an order book, which is the elementary mechanism of price formation, works. Then we move on to the description of the main statistical properties of order books and a brief review of their typical dynamics.
- **Chapter** 6: We introduce a zero-intelligence model for the order-book dynamics and focus our attention on the study of the impact of orders. The model is developed to investigate the role of liquidity in the price response to incoming orders. In order to perform such an analysis we introduce the concept of *price impact surface*, which is a two-dimensional function that permits one to study the response of the order book with respect to the volume of the orders and to the degree of liquidity of the order book. As a proxy for liquidity we introduce the granularity, which is a measure of the linear density of orders in the order book. We find that the dependence of the price response on the orders' volume is a power law with an exponent very close to the empirical findings revealed in the previous chapter, while the response is inversely dependent on the granularity. This implies that

when a liquidity crisis takes place and vanishing granularity is observed the price response is divergent.

- **Chapter** 7: Starting from the analysis of price time series of stocks traded at NYSE, we observe that odd spreads are more likely than even ones. The origin of this asymmetry is linked to the discrete nature of prices, in fact such an effect would not be observed if prices were continuous variables. However, we do not obtain a complete quantitative agreement between the empirical asymmetry and the predicted one. In fact we observe that there exists a second source of asymmetry, which reveals a strategic mechanism of placement of orders. We investigate the effect of these strategies in the framework of the model of the previous chapter.
- **Chapter** 8: This chapter is entirely devoted to the analysis of high-frequency data of the London Stock Exchange. We try to give a quantitative background to chartist strategies. In particular, we investigate whether there exists some evidence of memory effect induced by *special* values of the price (called supports and resistances) on which the price trajectory tends to bounce. We find that indeed such an effect is present and we observe that the more the price bounces on one of this level, the more the probability of observing a new bounce increases. This an example of self-fulfillment of a market belief through a self-reinforcement of the agents' perception of the strength of the resistance or the support.
- **Chapter** 9: Here we investigate an interesting regime of the relationship between skewness and kurtosis. In fact we find that datasets dominated by one large event are characterized by a peculiar relationship between skewness and kurtosis. We introduce a simple model that perfectly matches this special regime in the skewness–kurtosis plane. Finally we discuss the financial consequences of the asymmetric pattern observed in this plane and a possible quantitative method to determine the nature of the large events, i.e., whether they are really off statistics (outliers) or whether they are only rare events that are still compatible with the statistics given by the dataset (black swan).
- **Chapter** 10: In the last five years new data-driven research fields have gained great popularity: mathematical methods of computer science, statistical physics, and sociometry provide insights into a wide range of disciplines ranging from social science to human mobility. In this chapter we study whether non-financial data can be used as proxies for financial activity and whether this data can predict market movements. We find that query log volumes, i.e., the volume of users' searches on the Yahoo! search engine, are correlated with trading volumes and that, surprisingly, query log volumes anticipate trading volumes by one or two days.

Part III

- **Chapter** 11: In this short chapter we present the main paradigms developed by the classical theory of economic growth of countries. Over the last two centuries, the theories have been principally based on Adam Smith's work, according to which production specialization produces productive efficiency and consequently the greater the specialization of the productive system, the higher the wealth of a nation. Therefore one of the key elements of the classical theories of economic growth is the prediction of a highly specialized productive system. In the next two chapters we show instead that developed and rich countries produce almost all products and therefore a new paradigm is needed to grasp the complexity of economic growth.

- **Chapter** 12: Adam Smith's theory of specialization does not explain why the richest and most competitive countries have a diversified productive system that produces almost everything. A possible answer may be that the existence of some activities that emerge from the division of labour cannot be exported or traded but instead remain localized in the country where they have been developed. In this chapter we expose the economic theory of these non-tradable activities, called capabilities, developed in [35]. We also discuss the method of reflections introduced in this paper to measure the level of complexity of a productive system. While we accept the underlying economic theory we show that the method of reflection does not correctly capture the level of complexity of a country (i.e., the number of capabilities of a country).
- **Chapter** 13: The weak points of the method of reflections [35] are the starting point for developing a new iterative method to measure the competitiveness of a country and the complexity of a product by its quality. We introduce two variables, the country fitness and the product quality, whose asymptotic values allow the definition of a new non-monetary metric to measure the complexity of an economic system. This new metric on one hand measures the potential of development of a country's economy with respect to the current GDP level reached. On the other hand the values of the product quality represent a way to define and rank the products that surpasses the traditional and problematic definition of product qualities in terms of required hours of specialized labour.

References

1. Tacchella, A., Cristelli, M., Caldarelli, G., Gabrielli, A., & Pietronero, L. (2012). A new metrics for countries' fitness and products' complexity. *Scientific Reports on Nature, 2*, 723.
2. Cristelli, M., Tacchella, A., Caldarelli, G., Gabrielli, A., & Pietronero, L. (2013). Measuring the intangibles: a metrics for the complexity of countries and products. *PLoS ONE* (accepted).
3. Tacchella, A., Cristelli, M., Caldarelli, G., Gabrielli, A., & Pietronero, L. (2013). Economic complexity: conceptual grounding of a new metric for global competitiveness. *Journal of Economic Dynamics and Control*, *37*(8), 1683–1691.
4. Bouchaud, J.-P. (2008). Economics needs a scientific revolution. *Nature*, *455*, 1181.
5. Black, F., & Scholes, M. (1973). The pricing of options and corporate liabilities. *Journal of Political Economy*, *81*, 637.
6. Bouchaud, J.-P., Kockelkoren, J., & Potters, M. (2004). Random walks, liquidity molasses and critical response in financial markets. *Quantitative Finanace*, *6*, 115.
7. Lillo, F., & Farmer, J. (2004). The long memory of the efficient market. *Studies in Nonlinear Dynamics & Econometrics*, *8*, 1226.

8. Farmer, J. D., Gerig, A., Lillo, F., & Mike., S. (2006). Market efficiency and the long-memory of supply and demand: Is price impact variable and permanent or fixed and temporary. *Quantitative Finanace, 6*, 107.
9. Bouchaud, J.-P., Farmer, J. D., & Lillo, F. (2008). *How Markets Slowly Digest Changes in Supply and Demand*. Elsevier: Academic Press.
10. Chakraborti, A., Toke, I. M., Patriarca, M., & Abergel, F. (2011). Econophysics: Empirical facts and agent-based models. *Quantitative Finance*, (in press), http://arxiv.org/abs/arXiv:0909.1974v2.
11. Bouchaud, J.-P., Mezard, M., & Potters., M. (2002). Statistical properties of stock order books: empirical results and models. *Quantitative Finanace, 2*, 251.
12. Potters, M. & Bouchaud., J.-P. (2003). More statistical properties of order books and price impact. *Physica A, 324*, 133.
13. Wyart, M., Bouchaud, J.-P., Kockelkoren, J., Potters, M., & Vettorazzo., M. (2008). Relation between bid-ask spread, impact and volatility in order-driven markets. *Quantitative Finance, 8*, 41.
14. Lillo, F., Farmer, J. D., & Mantegna, R. N. (2002). Single curve collapse of price impact function for the New York Stock, Exchange. *arXiv:cond-mat/0207428*.
15. Farmer, J. D., Gillemot, L., Lillo, F., Mike, S., & Sen., A. (2004). What really causes large price changes? *Quantitative Finance, 4*, 383.
16. Lillo, F., Farmer, J. D., & Mantegna, R. N. (2003). Master curve for price-impact function. *Nature*, *421*, 129.
17. Weber, P., & Rosenow, B. (2006). Large stock price changes: volume and liquidity. *Quantitative Finance*, *6*, 7.
18. Cristelli, M., Zaccaria, A., and Pietronero, L. (2011). Critical overview of agent-based models for economics. In *Proceedings of the School of Physics E. Fermi, course CLXXVI, 2010*, Varenna. http://arxiv.org/abs/arXiv:1101.1847.
19. Alfi, V., Cristelli, M., Pietronero, L., & Zaccaria, A. (2009). Minimal agent based model for financial markets I: origin and self-organization of stylized facts. *European Physical Journal B*, *67*, 385. http://arxiv.org/abs/0808.3562.
20. Alfi, V., Cristelli, M., Pietronero, L., & Zaccaria, A. (2009). Minimal agent based model for financial markets II: statistical properties of the linear and multiplicative dynamics. *European Physical Journal B*, *67*, 399. http://arxiv.org/abs/0808.3565.
21. Alfi, V., Cristelli, M., Pietronero, L., & Zaccaria, A. (2009). Mechanisms of self-organization and finite size effects in a minimal agent based model. *Journal of Statistical Mechanics*, P03016.
22. Samanidou, E., Zschischang, E., Stauffer, D., & Lux, T. (2007). Microscopic models of financial markets. *Reports on Progress in Physics*, *70*, 409. doi:10.1088/0034-4885/70/3/R03
23. Lux, T., & Marchesi, M. (1999). Scaling and criticality in a stochastic multi-agent model of a financial market. *Nature*, *397*, 498.
24. Giardina, I., & Bouchaud, J.-P. (2003). Bubbles, crashes and intemittency in agent based market. *European Physical Journal B*, *31*, 421.
25. Caldarelli, G., Marsili, M., & Zhang, Y.-C. (1997). A prototype model of stock exchange. *European Physical Letter*, *40*, 479.
26. Alfi, V., Pietronero, L., & Zaccaria, A. (2009). *EPL*, *86*, 58003.
27. Battiston, S., Puliga, M., Kaushik, R., Tasca, P., & Caldarelli, G. (2012). Debtrank: Too central to fail? financial networks, the fed and systemic risk. *Scientific Reports on Nature*, *2*, 541.
28. Golub, T. (2010). Counterpoint: data first. *Nature*, *464*, 679.
29. Evans, J., & Rzhetsky, A. (2010). Machine science. *Science*, *329*, 399.
30. Lazer, D., et al. (2009). Life in the network: the coming age of computational social science. *Science*, *323*, 5915.
31. Gonzalez, M., Hidalgo, C., & Barabasi, A.-L. (2008). Understanding individual human mobility patterns. *Nature*, *453*, 479.
32. Choi, H. & Varian, H. (2009). Predicting the present with google trends. *Technical report*. http://static.googleusercontent.com/external_content/untrusted_dlcp/www.google.com/it//googleblogs/pdfs/google_predicting_the_present.pdf.

33. Bordino, I., Battiston, S., Caldarelli, G., Cristelli, M., Ukkonen, A., & Weber, I. (2012). Web search queries can predict stock market volumes. *PLoS ONE*, *7*, e40014.
34. Hidalgo, C. A., Klinger, B., Barabási, A. L., & Hausmann, R. (2007). The Product Space Conditions the Development of Nations. *Science*, *317*, 482.
35. Hidalgo, C. A., & Hausmann, R. (2009). The building blocks of economic complexity. *Proceedings of the National Academy of Sciences*, *106*, 10570.
36. Caldarelli, G., Cristelli, M., Gabrielli, A., Pietronero, L., Scala, A., & Tacchella, A. (2012). A network analysis of countries' export flows: firm grounds for the building blocks of the economy. *PLoS ONE*, *7*, e47278. http://arxiv.org/abs/arXiv:1108.2590.
37. Gaulier, G., & Zignago, S. (2010). Baci: International trade database at the product-level. Available from: http://www.cepii.fr/anglaisgraph/workpap/pdf/2010/wp2010-23.pdf.
38. Smith, A. (1776). In W. Strahan, & T. Cadell (Eds.), *The Wealth of Nations*, London: Onxofrd University Press.
39. Ricardo, D. (1817). *On the Principles of Political Economy and Taxation*. John Murray.
40. Romer, P. (1990). Endogenous technological change. *Journal of Political Economy*, *98*, 71.
41. Grossman, G. M., & Helpman, E. (1990). Quality ladders in the theory of growth. *Review of Economic Studies*, *98*, 71.
42. Flam, H., & Flanders, M. (1991). *Heckscher-Ohlin trade theory*. Cambridge, Mass.: MIT Press.
43. Cristelli, M., Alfi, V., Pietronero, L., & Zaccaria, A. (2010). Liquidity crisis, granularity of the order book and price fluctuations. *European Physical Journal B*, *73*, 41.
44. Zaccaria, A., Cristelli, M., Alfi, V., Ciulla, F., & Pietronero, L. (2010). Asymmetric statistics of order books: The role of discreteness and evidence for strategic order placement. *Physical Review E*, *81*, 066101.
45. Alfi, V., Cristelli, M., Pietronero, L., & Zaccaria, A. (2010). Reconstructing agents' strategies from price behavior. In Abergel et al. (Ed.) *Econophysics of Order-driven Markets*, Springer.
46. Preliminary findings regarding the market events of May 6, 2010. Technical report, SEC (2010). Available from: http://www.sec.gov/sec-cftc-prelimreport.pdf.
47. Garzarelli, F., Cristelli, M., Zaccaria, A., & Pietronero, L. (2011) Empirical evidence of technical trading. *Journal Statistical*, (submitted), http://arxiv.org/abs/1110.5197.

Chapter 2
Stylized Facts

The name Stylized Facts refers to all non trivial statistical evidences which are observed throughout financial markets. Almost all price time series of financial stocks and indexes approximatively exhibit the same statistical properties (at least qualitatively). In addition it has been shown that Stylized Facts are robust on different timescales and in different stock markets [1].

The systematic study of Stylized Facts has begun in very recent time (approximately from '90) for two reasons: a technical and a cultural one. The former one is that the huge amount of empirical data produced by financial markets are now easily available in electronic format and can be massively studied thanks to the growth of computational power in the last two decades. In order to make a comparison with some traditional fields of Physics, a similar quantity of information is observed only in the output of a big particle accelerator. The latter instead is due to the fact that traditional approaches to economic systems neglect empirical data as candidates respect to which a theory must be compared differently from Physics. From this point of view standard Economics is not an observational science.

Turning now our attention to the experimental evidences of financial markets, the main Stylized Facts are

- the absence of simple arbitrage,
- the power law decay of the tails of the return distribution,
- the volatility clustering.

In the following sections we analyze them.

2.1 Absence of Simple Arbitrage

The absence of simple arbitrage in financial markets means that, given the price time series up to now, the sign of the next price variation is unpredictable on average. In other words it is impossible to make profit without dealing with a risky investment. This implies that the market can be seen as an open system which continuously reacts to the interaction with the world (i.e. trading activity, flux of information, etc) and

M. Cristelli, *Complexity in Financial Markets*,
Springer Theses, DOI: 10.1007/978-3-319-00723-6_2,

self-organizes in order to quickly eliminate arbitrage opportunities. This property is also called arbitrage efficiency.

This condition is usually equivalent to the informational efficiency expressed in economic literature saying that the process described by the price p_t is a martingale that is

$$E[p_t|p_s] = p_s \tag{2.1}$$

where $t > s$. Here we are assuming that the price is a synthetic variable which reflects all the information available at time t. If this is not true the conditioning quantity is the available information I_s at time s and not only the price p_s.

However, the condition of martingale is uneasy from a practical point of view and the two-point autocorrelation function of returns is usually assumed as a good measure of the market efficiency

$$\rho(\tau, t) = \frac{E[r_t r_{t+\tau}] - E[r_t]E[r_{t+\tau}]}{E[r_t^2] - E[r_t]^2}. \tag{2.2}$$

If the process $\{r_t\}$ is at least weakly stationary then Eq. 2.2 simply becomes $\rho(\tau) = (E[r_t r_{t+\tau}] - \mu_r^2)/\sigma_r^2$ where $\mu_r = E[r_t]$ and $\sigma_r = E[r_t^2] - E[r_t]^2$. If the autocorrelation function of returns is always zero we can conclude that the market is efficient.

In real markets the autocorrelation function is indeed always zero (see Fig. 2.1) except for very short times (from few seconds to some minutes) where the correlation is negative (see inset of Fig. 2.1). The origin of this small anti-correlation is well-known and due to the so-called *bid ask bounce*. This is a technical reason deriving from the double auction system which rules the order book dynamics (see [2] for further details).

In the end we want to stress that the efficiency is a property that holds on average: locally some arbitrage opportunities can appear but, as they have been exploited, the efficiency is restored [3, 4].

2.2 Fat-Tailed Distribution of Returns

The distribution of price variations (called returns) is not a Gaussian and prices do not follow a simple random walk. In details very large fluctuations are much more likely in stock market with respect to a random walk and dramatic crashes are approximately observed every 5–10 years on average. These large events cannot be explained by gaussian returns. Therefore to characterize the probability of these events we introduce the complementary cumulative distribution function $F(x)$

$$F(x) = 1 - \text{Prob}(X < x) \tag{2.3}$$

which describes the tail behavior of the distribution $P(x)$ of returns.

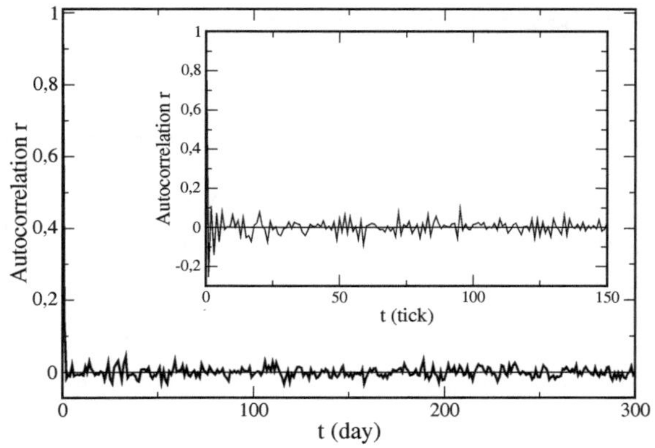

Fig. 2.1 We report the autocorrelation function of returns for two time series. The series of the main plot is the return series of a stock of New York Stock Exchange (NYSE) from 1966 to 1998 while the series of the inset is the return series of a day of trading of a stock of London Stock Exchange (LSE). As we can see the sign of prices are unpredictable that is the correlation of returns is zero everywhere. The time unit of the inset is the tick, this means that we are studying the time series in event time and not in physical time

The complementary cumulative distribution function $F(x)$ of real returns is found to be approximately a power law $F(x) \sim x^{-\alpha}$ with exponent in the range 2–4 [5], i.e. the tails of the probability density function (pdf) decay with an exponent $\alpha + 1$. Since the decay is much slower than a gaussian this evidence is called Fat or Heavy Tails. Sometimes a distribution with power law tails is called a Pareto distribution. The right tail (positive returns) is usually characterized by a different exponent with respect to the left tail (negative returns). This implies that the distribution is asymmetric in respect of the mean that is the left tail is heavier than the right one ($\alpha^+ > \alpha^-$).

Moreover the return pdf is a function characterized by positive excess kurtosis, a Gaussian being characterized by zero excess kurtosis. In Fig. 2.2 we report the complementary cumulative distribution function $F(x)$ of real returns compared with a pure power law decay with exponent $\alpha = 4$ and with a gaussian with the same variance.

When the tail behavior of the return distribution is studied varying the time lag at which returns are performed [1], a transition to a gaussian shape is observed for yearly returns. However it is unclear if this transition is genuine or due to a lack of statistics or to the non stationary return time series.

2.3 Volatility Clustering

In the lower panel of Fig. 2.3 we report the return time series of a NYSE stock (returns are here defined as $\log(p_{t+1}/p_t)$). As we can see the behavior of returns appears to be intermittent in the sense that periods of large fluctuations tend to be followed by

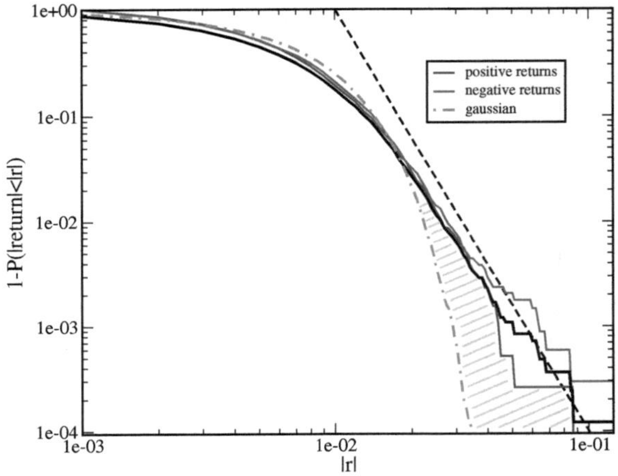

Fig. 2.2 We report the complementary cumulative distribution function of the absolute value of returns (*solid black line*). The *green dashed line* (·—) is the complementary cumulative distribution function of a gaussian with the same variance of the real return distribution. The *dashed black line* is a pure power law decay with exponent $\alpha = 4$. The *blue* and *red lines* are instead the complementary cumulative distribution functions for positive and negative returns respectively. We can see that *red curve* has a slower decay with respect to the *blue* one. This asymmetry between positive and negative returns is the origin of the non zero skewness of the probability density function of returns

large fluctuations regardless of the sign and the same behavior happens for small ones.

In Economics the magnitude of price fluctuations is usually called volatility. It is worth noticing that a clustered volatility does not deny the fact that returns are uncorrelated (i.e. arbitrage efficiency). Therefore the magnitude of the next price fluctuations is correlated with the present one while the sign is still unpredictable. In other words stock prices define a stochastic process where the increments are uncorrelated but not independent.

Different proxies for the volatility can be adopted: widespread measures are the absolute value and the square of returns. As a consequence of the previous considerations about the clustering of volatility, the autocorrelation function of absolute (or square) returns is non zero. We also find that the autocorrelation is well-described by a power law decay with exponent ranging from -1 to 0 as reported in Fig. 2.4. The very slow decay means that volatility is correlated on very long time scales from minutes to several months/years. The exponent of the autocorrelation function is not universal as the one of fat tails but it is typically around 0.2–0.3. The volatility clustering was observed the first time by Mandelbrot in 1963 [6].

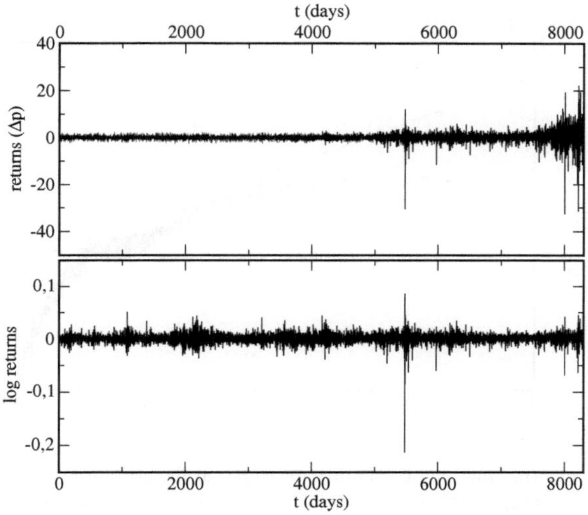

Fig. 2.3 Return time series of a stock of NYSE from 1966 to 1998. The two figures represent the same price pattern but returns are differently computed. In the *top* figure returns are calculated as simple difference, i.e. $r_t = p_t - p_{t-\Delta t}$ while in the *bottom* one returns are log returns that is $r_t = \log p_t - \log p_{t-\Delta t}$. From the lower plot we can see that volatility appears to be clustered and therefore large fluctuations tend to be followed by large ones and vice versa. The visual impression that the return time series appears to be stationary for log returns suggests the idea that real prices follow a multiplicative stochastic process rather than a linear process

2.4 Other Stylized Facts

Beyond these Stylized Facts we can state other relevant effects which are widespread in financial markets such as

- the gain/loss asymmetry, i.e. one observes large drawdowns in stock prices and stock index values but not equally large upward movements. This is linked to the asymmetry of the return pdf.
- leverage effect: the volatility of an asset are negatively correlated with the returns of that asset.
- trading volume and volatility are correlated.

See also [1, 3–5, 7, 8] for more details about Stylized Facts and their analysis.

2.5 Stationarity and Time-Scales

Before turning our attention to the analysis of the models which try to interpret Stylized Facts, we want to discuss a final question: the stationarity and the time scales of the observation of financial markets.

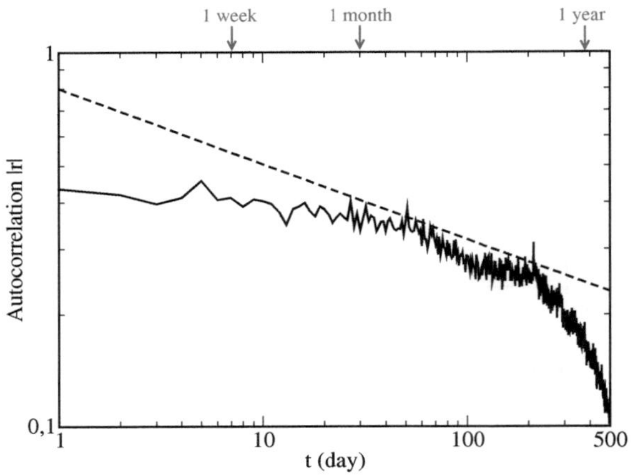

Fig. 2.4 Autocorrelation function of volatility measured as the absolute value of returns. We find that the function can be approximately described by a power law. We report a pure power law decay with exponent 0.2 for comparison. The return time series used in this analysis is the previous NYSE one from 1966 to 1988. It is worth noticing that the volatility is significantly correlated and then clustered on time scale longer than one year

The hypothesis of stationarity is usually invoked because, if satisfied, the statistical properties of the phenomenon under consideration become invariant under temporal translation. In the case of financial markets it is not clear whether the return time series verifies this condition: intraday activities, seasonality, weekends, holidays, Economy growth are elements that can *a priori* make the returns not stationary. However, it can be argued that the process is observed on the wrong scale (see [9]) and on a larger time scale the process may become stationary.

Last but not least it has been hypothesized that the non stationarity of financial series derives from the fact that they are studied in physical time units. On this account it has been proposed to define a time rescaling such that the transformation which makes stationary the financial data is the correct one. However, the choice of which elements should be involved in this transformation is arbitrary and ranges form seasonality of the calendar to the volumes of trading activity. The question is still open (see [10–13] for further references).

References

1. Cont, R. (2001). Empirical properties of asset returns: stylized facts and statistical issues. *Quantitative Finance*, *1*, 223.
2. Harris, L. (2003). *Trading and exchanges*. Oxford: Oxford University Press.
3. Mizuno, T., Takayasu, H., & Takayasu, M. (2007). Analysis of price diffusion in financial markets using PUCK model. *Physica A*, *382*, 187. http://arxiv.org/abs/arXiv:physics/0608115.

4. Takayasu, M., Mizuno, T., & Takayasu, H. (2006). Potential force observed in market dynamics. *Physica A*, *370*, 91.
5. Bouchaud, J.-P. & Potters, M. (2003). *Theory of financial risk and derivative pricing: from statistical physics to risk management*. Cambridge: Cambridge University Press.
6. Mandelbrot, B. (1963). The variation of certain speculative prices. *Journal of Business*, *35*, 394.
7. Mantegna, R. N., & Stanley, H. (2000). *An introduction to econophysics: correlation and complexity in finance*. USA: Cambridge University Press.
8. Chakraborti, A., Toke, I. M., Patriarca, M., & Abergel, F. (2011). Econophysics: Empirical facts and agent-based models. *Quantitative Finance*, (in press). arXiv: http://arxiv.org/abs/arXiv:0909.1974v2.
9. LeBaron, B. (2001). Stochastic volatility as a simple generator of apparent financial power laws and long memory. *Quantitative Finance*, *1*, 621.
10. Andersen, T. G., & Bollerslev, T. (1998). Intraday periodicity and volatility persistence in financial markets. *Journal of Empirical Finance*, *4*, 115.
11. Ané, T., & Geman, H. (1999). Stochastic volatility and transaction time: an activity based volatility estimator. *Journal of Risk*, *2*, 57.
12. Bak, P., Tang, C., & Wiesenfeld, K. (1987). Self-organized criticality: An explanation of the 1/f noise. *Physical Review Letter*, *59*, 4.
13. Mizuno, T., Takayasu, M., & Takayasu, H. (2005). Modeling a foreign exchange rate using moving average of yen-dollar market data. http://arxiv.org/pdf/physics/0508162.

Part I
Agent-Based Modeling

Chapter 3
Critical Review of Agent-Based Models

3.1 Introduction

In this chapter we present a critical discussion of some of the most representative Agent-Based Models (ABM hereafter) of Economics. We embed them in a framework which highlights their key principles and the relative strengths and weaknesses. The objective is to point out the possible lines of convergency and improvement in order to focus towards an optimal model. A crucial weakness in this respect is that the experimental framework defined by the so-called Stylized Facts is still rather limited and an improvement of this body of knowledge appears to be the bottle neck in the field.

3.1.1 Classical Theory of Economics

The Classical theory of Economics is based on the following elements [1, 2]

- Situation of equilibrium with equal (representative) agents which are (quasi) rational, have the same information and process it in the same way;
- Important price changes correspond to new information which arrives on the market. The fact that this information is random and independent leads to the famous random walk model [3, 4] and the corresponding Black and Scholes equations [5, 6];
- This new information modifies the ratio between offer and demand and then also the price. This corresponds to a mechanical equilibrium of the market;
- These concepts also imply a Cause-Effect relation in which large price changes are due to the market reaction to the arrival of exogenous important news. Therefore a large price change is supposed to be associated to an equally large exogenous event.

It is too easy to argue that most of these assumptions have no basis at all. So one may wonder why they are so widely adopted. The real reason is that they permit an analytical treatment of the problem. This is a very different perspective from natural

M. Cristelli, *Complexity in Financial Markets*,
Springer Theses, DOI: 10.1007/978-3-319-00723-6_3,

Fig. 3.1 This cartoon appeared on the cover of The Economist on November 7, 1997 to illustrate non rationality, limited information and a strong herding behavior. All these elements are absent in the standard model of Economics. The ambition of the agent-based models is to include this type of elements in the modeling of Economics

sciences in which very few realistic problems can be treated exactly but still one can get an appreciable level of understanding with suitable approximation schemes or computer simulations. In addition these ideas are usually not tested against empirical data; a fact that strongly limits the scientific basis of the whole framework.

The most reasonable of the above assumptions is the fact that indeed many external news are random and incoherent so the random walk appears reasonable as a simple modeling. On the other hand the assumptions made for the behavior of the agents are very far from reality. A famous cartoon of the Economist to stress this point is reproduced in Fig. 3.1. The agents can be very different from each other, their level of information and the way the use it is also very different. Finally they are not at all independent and, especially in situations of crisis the rationality hypothesis can be seriously wrong. In these situations fear and panic, as well as euphoria, lead to very strong herding behavior (Fig. 3.1) which is completely neglected in the standard model.

From an empirical point of view one can observe marked deviations from the standard picture:

- Great catastrophic events like the '87 crash, the Internet bubble of 2000 and the recent case of the subprimes do not seem to have any relation with specific events or new information;
- There are clear deviations from the gaussian behavior for the price distribution which are named Stylized Facts. These can be observed for the large price fluctuations but are actually present at all scales, even in relatively calm periods;
- Breaking of the Cause-Effect relation. Often large price fluctuations are not associated to any special event. In Table 3.1 we report the top ten price movements from the second world war til 1989 together with their origin as reported by the media. Even for the recent subprime crisis the total amount of money corresponding to the subprimes is very small with respect to the catastrophyc effect it had worldwide.

3.1.2 Towards Complexity

This situation naturally calls for a possible description in terms of critical phenomena and complex systems [8]. The study of complex systems refers to the emergence of collective properties in systems with a large number of parts in interaction among them. These elements can be atoms or macromolecules in a physical or biological context, but also people, machines or companies in a socio-economic context. The

Table 3.1 We report the ten largest price movement in the Standard & Poor's index from the second world war till 1989 together with the official explanation of their "origin" as reported in the media (see [7] for further details)

Date	Percent change	New York times explanation
Oct. 19, 1987	−20.47	Worry over dollar decline and trade deficit
Oct. 21, 1987	+9.10	Interest rates continue to fall, bargain hunting
Oct. 26, 1987	−8.28	Fear of budget deficit, margin calls
Sep. 3, 1946	−6.73	"... No basic reason for the assault on prices"
May 28, 1962	−6.68	Kennedy forces rollback of steel price hike
Sep. 26, 1955	−6.62	Eisenhower suffers heart attack
Jun. 26, 1950	−5.38	Outbreak of Korean War
Oct. 20, 1987	+5.33	Investors looking for "quality stocks"
Sep. 9, 1946	−5.24	Labor unrest in maritime and tracking industries
Oct. 16, 1987	−5.16	Fear of trade deficit and tension with Iran

science of complexity tries to discover the nature of the emerging behavior of complex systems, often invisible to the traditional approach, by focusing on the structure of the interconnections and the general architecture of systems, rather than on the individual components. This implies a change of paradigm form the previous mechanical model to a complex model in which intrinsical instabilities can develop in a self-organized way without a cause-effect relation [1, 8–10].

3.1.3 Why Agent-Based Models

The so-called ABM represent a broad class of models which have been introduced to describe the economic dynamics in a more realistic way. Their building blocks are:

- The agents are heterogeneous with respect to their various properties like strategies, wealth, time scale, etc;
- The interaction between them is a fundamental element and, of course, it can have many different characteristics;
- Price dynamics depends on the balance between offer and demand but the specific implementation can be different.

Usually the dynamics of this type of models is able to produce deviations from the gaussian behavior related to the Stylized Facts for some specific range of parameters. Additional elements which are often neglected but are important in our opinion are the fact that the total number of agents or strategies can strongly vary in time and that the system has to evolve spontaneously towards the quasi-critical state which shows the Stylized Facts.

3.2 Standard Economic Theory

In the previous section we have reported the main empirical features of financial markets. Now we briefly review the so-called neoclassical framework which has been the dominant paradigm of economic theory for fifty years.

According to standard economic theories markets are in an equilibrium state.

Before analyzing the consequence of such an assumption, let us briefly describe how an economy is modeled by the neoclassical approach. The economic system is composed of goods, agents and firms. Agents and firms own a certain amount of goods, have a set of technologies to transform goods in other goods and an utility function quantifies agents and firms' preferences in good consumption. Agents and firms only aim at maximizing their utility by selling, buying, producing or consuming goods.

The typical hypothesis of equilibrium models are:

1. A single agent's action is marginal with respect to the price, that is the price variations due to the single agent is a contribution of higher order with respect to the price variation;
2. Agent's aim is the maximization of the utility given the price vector of goods;
3. The price vector is the solution of the equation that makes the aggregate demand and the aggregate supply equal;
4. Agents are fully rational and are fully informed, that is they have perfect models of reality to determine their actions and know the price vector.

There are two main motivations of the great success of equilibrium models.

The former is that equilibrium models are a closed and relatively simple quantitative theory for economic system. The models are closed in the sense that given the previous four assumptions no extra ingredients are needed to determine the equation connecting goods, technology, production, trade and prices.

The second reason is that if an extra hypothesis is added to the behavior of the utility function and of technology, then the equilibrium state always exists [11]. The demonstration of the existence of this equilibrium state consists in the demonstration of the existence of a fixed point of a suitable map acting on the space of price vectors. The two additional hypothesis can be summarized saying that the more an agent consumes a good, the less the agent will obtain a benefit from the consumption of an extra unit of that good and that the more a technology is used to produce a good, the less the extra production derives from one extra unit of input. We do not report the problems that arise from the question of how this fixed point is reached. We only highlight that the existence of this point does not guarantee that the system is always able to reach this state.

The models herein proposed are essentially one-time models. In their original formulation they are static and do not deal with the dynamics of the price evolution and of agent's allocation of goods. However, it has been shown that equilibrium models can be extended to a multi-time set up and even to an economy composed of financial goods and not only of concrete and physical goods. The details of these two features go beyond the goal of the present review. The readers who are interested in these questions can find more information in [2, 12, 13].

Now we want to analyze the consequence of the equilibrium assumptions in a stock market. Because of the multi-time horizon we have to introduce a new hypothesis and we are going to see that it is the stronger assumption made until this point. In a multi-time set up, agent's rationality implies the knowledge of all the present prices but also of all the dependences of prices on the possible future states on which the agents are uncertain (these states are called states of Nature). Furthermore the extension from a one-time to a multi-time horizon implies that the maximization of the utility function must now depends on all the possible future scenarios. Thus, whether or not a good is consumed today depends on all the possible future history of price vector and on all the probabilities of the uncertain states of Nature. Thus the consequence of the full rationality become a bit strange with respect to our daily experience about preferences in uncertain conditions. In particular in equilibrium models we have to assume that the probability of the uncertain states of Nature are known or perfectly deduced by rational agents at the beginning of the history. This also implies that the uncertain quantities are stationary in time and on all time scales.

On the other end it is simple to show that in the original one-time equilibrium models the equilibrium state is also a Pareto efficient state. From this point of view the model is allocative efficient. In fact a state is defined Pareto efficient if there is not any change of the allocation of all agents that make at least one agent happier with respect to his/her utility. The extension to financial markets is far from being trivial because the equilibrium state in a multi-time horizon and in uncertain conditions is not usually a Pareto efficient state. Therefore rationality can be extended to financial contest even if it includes the unrealistic perfect knowledge of all future sources of uncertainty while efficiency is in general not satisfied by equilibrium model of financial markets.

For this reason a weaker efficiency is usually required to hold in financial contest: the informational efficiency which we have introduced in the previous section. In mathematical terms the informational efficiency means that the price process is a martingale with respect to the information process I_t. In its stronger version the price is a synthetic variable which incorporates all the information and therefore present price is the best prediction of any future price.

The most striking consequence of the full rationality assumption and of perfect informational efficiency is that prices should follow the fundamental value of the financial good that is the future discounted forecast of the dividend payed by the stock. In this market price and fundamental price would always coincide (see the brief argument in [14] to illustrate this point). We want to stress that all agents are fully rational and endowed with the same information, the logical consequence is that all agents, being rational, get the same forecast about future. Thus agents are rational and homogeneous and this permit to deductively forecast the future prices of a stock. This procedure leads to the formula

$$p_t = \sum_{k=1}^{\infty} \beta^k E[d_{t+k}|I_t] \tag{3.1}$$

where β is the discount factor and is equal to $1/(1+r)$ where r is the risk free rate (i.e. treasure bonds) and d_t is the amount paid out in dividends at time t.

We can see that the only allowed fluctuations of prices derive from the uncertainty due to the conditioning of the present information I_t. So that we have found prices will follow the fundamental value except for small deviations from it due to fluctuations of information. Moreover in such a context there are not any incentives to trade since there are not any speculative opportunities. The investors will gain profit only from the revenues of the dividends.

If we compare the predictions of standard neoclassical models with the observational results presented in the previous section we find that Stylized Facts can be hardly fitted by the framework of equilibrium theory. Even worse, equilibrium theory does not appear as the natural starting point for a new theory. Instead a paradigm shift is required and ABM are the first step in this direction. Actually, as equilibrium models, ABM are more similar to metaphors than to real predictive/quantitative models but their advantage relies in the fact that Stylized Facts are their starting point and their benchmark to determine their fitness in describing financial markets.

3.3 Agent-Based Models

In this section we present and discuss six ABM which well represent the development of this field. Differently from most reviews of this area [15–17], the analysis of each model is performed in a very schematic way according to some categories. These categories are:

- **Aim of the model:** we describe the reasons of the introduction of each model.
- **Agents and strategies:** we give a detailed description of the type of agents involved in the model, the strategies used to trade, the mechanism which rules the change of strategy (in the model where this feature is present), etc.
- **Number of agents:** we discuss if the number of active agents changes in time or is kept fixed.
- **Price formation:** in this section we discuss the mechanisms through which price evolution and strategy evolution interacts
- **Origin of Stylized Facts:** we analyze which elements give origin to Stylized Facts.
- **Discussion realism/tractability:** the development of an ABM sets a not trivial problem with respect to the compromise between tractability and realism. We review the advantages and disadvantages of each model in relation to this question.
- **Self-Organized Criticality**: if it is possible, we analyze how the problem of the market self-organization in the regime of Stylized Facts is addressed.

3.3.1 Kim and Markowitz Model: An Attempt to Explain the Black Monday

Aim

On Monday, October 19, 1987 stock markets worldwide underwent the largest draw-down of financial history in a very short time. In one day Dow Jones Industrial Average lost 508 points which correspond to a percentage drop of about 23 %.

The origins of this crash are still unclear and a conclusive explanation is still missing [18, 19].

The Black Monday's causes immediately appeared and in some extent still appear rather mysterious because there was not any flux of information which could explain the origin of such a large and such a widespread negative fluctuation.

The Black Monday had the merit to reveal to the world the inadequacy of the cause-effect relation between information and price fluctuations.

The research of the real causes beyond an anomalous flux of information was a perfect playground for the first modern multi ABM proposed by Kim and Markowitz in 1989 [20].

Agents and Strategies

The model is very focused on the causes of the Black Monday. In particular the authors of this model want to investigate whether the presence of hedging and portfolio insurance strategies [21] can destabilize the market or at least produce a significant increase of volatility. In fact at the very end the model tries to understand the link among volatility and portfolio insurance strategies.

In practice the model has two classes of agents. The former type is called *rebalancer*, the latter *portfolio insurer* and both classes can trade two assets. As in many ABM these two assets are a risk free asset and a risky one of which the return can be either positive or negative.

The first class operates in order to keep the proportion invested in cash and in the risky stock constant given the constraint of finite wealth. The rule to define the portfolio composition of portfolio insurers is more sophisticated. This kind of agents aims at keeping constant the following proportion, given the constraint deriving from finite wealth,

$$\frac{q_t p_t}{w_t - f} = \beta \tag{3.2}$$

where f is a fraction γ of the initial capital w_0 and $\beta > 1$, q_t the amount of the risky asset owned by the agent and p_t is the price of the risky asset. This time the proportion is performed with respect to the wealth of the agents decreased by the minimum amount f of the capital which these agents want to have guaranteed.

The rebalancers have a stabilizing effect on the price behavior. Instead portfolio insurers can destabilize the market because their response to a price variation is the opposite with respect to rebalancer action.

The model is characterized by heterogeneous strategies (i.e. two ones) but there is not a real heterogeneity inside the two types of agents because all rebalancers and all portfolio insurers respectively share the same parameters for their strategies).

Changes of strategies are not allowed in this model. The unique change consists in removing those agents whose wealth is zero.

Number of Agents

The number of agents is trivially dependent on the time because if the wealth of an agent goes to zero then she is removed from the market. Consequently the number of players is always non increasing. We said that this dependence is trivial because this model does not aim at investigating this feature and therefore it substantially belongs to the class of ABM characterized by a fixed number of agents.

Price Formation

The price evolves according to the imbalance between supply and demand due to the updates of the portfolio compositions. The mechanism for the execution of orders resembles a very simple order book-like mechanism which permits to take into account the effect of finite liquidity. Orders due to the review of the portfolio composition are executed only when a buying request is matched by a selling request at the same price.

A trading day is declared closed when every investors has reviewed their portfolio at least one time and the agents whose wealth w_t is equal to zero at the closing price are removed from the simulation. In [16] it has been shown that this rather complicated mechanism for the price update can be replaced by a more easier relationship among price variation and excess demand and the authors of [16] find approximately the same results.

Origin of Stylized Facts

This model does not directly deal with the problem of Stylized Facts as we have already mentioned. Some later studies (for instance see [16]) check whether the model is able to reproduce some Stylized Facts. The conclusion is negative but we cannot say that this is really a negative point because we are testing the model on a task for which it is not designed.

Discussion Realism/Tractability

The model was introduced to study the stability of financial markets with respect to agent strategies. Thus, being the first real ABM, the model has the merit for searching endogenous causes of the instabilities of markets. From this point of view this work is pioneering and opens a novel vision and approach of market problem. Nevertheless the absence of replacements for the bankrupt agents and the quasi-periodic pattern of prices reduces its success in explaining market instabilities.

Furthermore the model makes a set of too simplistic assumptions and introduces an unnecessary degree of complication for the mechanism of price evolution. In [16] the authors show that similar results can be obtained with a much easier price rule.

Self-Organized Criticality

This aspect is not addressed in this model.

3.3.2 Santa Fe Artificial Stock Market

Aim

The Santa Fe artificial stock market is the result of the collaboration of two computer scientists, a physicists and two economists at Santa Fe Institute [14]. In this section we will analyze the model presented in [14] which is the most interesting version with respect to Stylized Facts and complexity of financial markets (see also [22]). The work proposed in [14] was born in the same line of Kim and Markowitz model, the Black Monday and the consequent unexplainable worldwide crash in 1987.

The question of whether the market is efficient and rational or not is not only an interesting academic problem. In fact the conclusion of standard financial markets is that homogeneous rational agents endowed with the same information have no incentive to trade and the exchanged volumes should be nearly zero. In addition technical trading strategies cannot be profitable in an efficient world where prices always follow their fundamental value.

On this account a very interesting discussion about the main differences between Economics and Physics with respect to their approach to empirical evidences can be found in [23]. It is clearly reported that there is a systematic lack of empirical validation of economic models until very recent time.

This model aims at investigating the regime in which a market tends when the agents' forecasting strategies are heterogeneous. It is worth noticing that the authors maintain a neoclassical framework to model the market and the only new ingredient is indeed the heterogeneity in agents' forecasting action. Consequently the agents' homogeneous expectation does not hold and the price equilibrium problem cannot be anymore solved by deduction or by logical reasoning. In such a heterogeneous scenario this work wants to investigate whether or not the agents' strategies tend to a rational and homogeneous state or whether agents are able to select technical strategies and build a realistic dynamics of prices.

Agents and Strategies

This artificial market is composed of two tradable financial assets, a risk-free bond with interest rate r and a risky asset which pays a stochastic dividend d_t.

The set-up of the model is neoclassical, i.e. the model is built following the framework of standard financial theory. Differently from other ABM this model has a microeconomic foundation in some extent. In fact agents aim at maximizing their utility function $U(c)$ where c is their wealth. The maximization of the utility function corresponds to find the optimal composition of the portfolio between the bond and the risky stock. It is a well-known result of standard economic theory that the optimal allocation $x_{i,t}$ of the risky asset in such a scenario is

$$x_{i,t} = \frac{E_{i,t}[p_{t+1} + d_{t+1}] - p_t(1 + r)}{\lambda \sigma^2_{i,t,p+d}} \tag{3.3}$$

where $E_{i,t}[\cdot]$ is the forecast function used by agent i at time t and $\sigma^2_{i,t,p+d}$ measures the confidence of the prediction $E_{i,t}[p_{t+1} + d_{t+1}]$.

The deviation from the standard neoclassical scenario is introduced at this step: the rationality and homogeneity hypothesis would be normally invoked, so that the agents' index can be neglected and strategies confidences are necessarily all equal. The novelty of the model is the introduction of the heterogeneity in the agents' expectations $E_{i,t}[\cdot]$. Now it is impossible to solve the clearing price equation by deduction since agents must guess the others' expectation rules and there are not any logical criterions to perform such a task. Then agents inductively learn, they have to compete in a heterogenous world of strategies and form their expectations by adaptation.

Number of Agents

The number of active agents is constant over the time. The typical number of agents in [14] is 25. We are not aware of systematic study of the model properties with respect to the number of agents.

Price Formation

As in most ABM the feedback between agents' actions and prices is represented by the price clearing mechanism. The clearing price p_t at time t is the equilibrium price from the direct solution of the equation

$$\sum_{i=1}^{N} x_{i,t} = N \tag{3.4}$$

where i is the agent's index, $x_{i,t}$ is the demand of shares of the agent i at time t ($x_{i,t}$ can be both positive and negative) and N is the available number of shares.

Origin of Stylized Facts

The model is able to reproduce both the regimes it was introduced for depending on the values of the parameter that rules the frequency of activation of the genetic algorithm.

If this frequency is high enough the model reproduces a non rational regime—with respect to the concept of rationality in standard financial theories—otherwise the agents converge to a homogeneous strategy very close to the rational solution of this neoclassical artificial market and the price behavior follows the fundamental and small trade volumes are observed.

Discussion Realism/Tractability

This is one of the few model with a microeconomic foundation and a set-up very close to the traditional approach to financial markets.

However, by simply introducing the heterogeneity in the future expectations of agents the model shows that non-rational scenario dominated by technical traders can become the inductive *rational* solution, where by rational we mean that the technical strategies in such a regime become competitive with respect to fundamental ones and are not eliminated from the market.

The model does not allow for any analytical approach and has a very limited tractability.

Self-Organized Criticality

The model shows that there exist two stable states which are mainly driven by the frequency of activation of the genetic algorithm. One of these two states produces a price dynamics which has statistical properties which are very similar to the real ones. Thus, as many other ABM, it shows that a stable and self-regulated state which exhibits Stylized Facts can exist but the model does not address the problem of how this state can be reached and why markets should self-organize around this values of the frequency for the genetic algorithm.

In brief the model does not cope with the dynamics of the market self-organization.

3.3.3 Minority Game

Aim

Thanks to its mathematical structure, the Minority Game (MG) [24, 25] is the most studied ABM. Consequently, many different versions of this model exist, with very different aims. The very first formulation, due to Arthur [26], is known as "The El Farol Bar Problem" and studies inductive reasoning in scenarios of bounded rationality.

Challet and Zhang [27] gave a mathematical formulation of the problem. In this review we will analyze a model proposed by Challet et al. [28] that specifically addresses the problem of the appearance of the Stylized Facts, still conserving all the main ingredients of the classic MG.

Agents and Strategies

The MG is based on a simple concept. Let us suppose that an agent can choose between two options (these options are typically buy or sell a given amount of stocks). That agent will receive her payoff if, after all the agents choice, it will turn out that she took the minority side, that is she pick the less chosen option. This criterion may appear simplistic, but gets the point that one can exploit the price variations due to other agents' operations if one has previously taken the opposite side.

We now discuss the generic features of the MG.

There are N agents in the market. At each time step t each agent can choose to buy or sell stocks. Consequently, the function that describes the action of the ith agent, $a_i(t)$, can be equal to either $+1$ or -1. We define the excess demand at time t as

$$A(t) = \sum_{i=1}^{N} a_i(t). \tag{3.5}$$

It should be noticed that the agent can not choose the volume of her operation.

The fact that the minority side wins is considered introducing an utility function or a payoff given by

$$U_i(t) = -a_i(t)\, F[A(t)]$$

where F is an odd function of $A(t)$.

Number of a Agents

As we have seen in the previous section, the total number of agents is fixed, but at each time step they can choose to operate or not, so that the effective number of active agents varies over time. For this reason, using an analogy with Statistical Mechanics, one often refers to this model as the Grand Canonical Minority Game (GCMG).

Price Formation

Being the excess demand $A(t)$ given by Eq. 3.5, the price is updated with the rule

$$\log p(t+1) = \log p(t) + \frac{A(t)}{\lambda}$$

where λ is a measure of the market depth. The return $r(t)$ is therefore basically given by the excess demand.

Stylized Facts and Their Origin

Volatility clustering, fat tails and market crashes display a power law behavior and increase with the normalized number of speculators, $n_s = N_s/P$. The exponents depend on the specific parameters used in the simulation. The responsible for large price movements are the inactive agents that does not interact when out of the market but can enter at the same time if their strategies are similar.

Challet et al. [29] studied a version of the GCMG that is very close to the one presented here. They found that in their model the Stylized Facts are finite size effects, being present in a limited region of the parameter space that shrinks as $N^{-1/4}$ in the limit $N \to \infty$.

Realism and Tractability

The level of realism and tractability is different for different versions of the MG.

Many realistic features have been added and studied taking the classic MG as a starting point: agents with finite or evolving wealth, agents using correlated strategies or acting at different frequencies, markets with more than one asset and so on (for these and other possible extensions, see [24, 25] and references therein). If one considers all versions, many different aspects of real financial markets are taken in consideration.

The MG is, among all the ABM, the most studied from an analytical point of view. The first step is due to Challet et al. [30], that showed that the stationary state of the dynamics is related to the minima of the so called *predictability*

$$H = \overline{\langle A \rangle^2} = \frac{1}{P} \sum_{\mu=1}^{P} \langle A|\mu \rangle^2.$$

This observation opened a connection with Statistical Mechanics, as it turned out that the ground state was the same of a system of disordered spins. For example, using the replica method [31] they solved the basic model exactly and showed that the ground state is not a Nash equilibrium [2], that is, it is not individually optimal. This kind of approach can give revealing insights into almost every version of the MG.

Self-Organized Criticality

The problem of Self-Organization is mentioned but not systematically addressed.

3.3.4 Caldarelli, Marsili and Zhang: A Prototype of Stock Exchange

Aim

In the work [32] the authors focus their attention on the endogenous mechanisms of financial markets and show that these ingredients are sufficient to obtain a stable and self-organized market in the sense that external (i.e. exogenous) rules are not required to stabilize the price behavior.

Agents and Strategies

There are two tradable assets: a risk free asset with zero-interest rate (i.e. cash) and a risky stock. All agents are speculators so that they all aim at increasing their wealth and at each time their action consists in updating their portfolio composition of stocks and cash according to their current strategy.

All the agents share the same information which consists in the price time series.The agents' strategies depend only on the price history and consists in updating her owned shares in the following way

$$\Delta S_{i,t} = X_{i,t} S_{i,t} + \frac{\gamma_i M_{i,t} - p_t S_{i,t}}{2\tau_i} \tag{3.6}$$

where the index i denotes the agents. The parameters γ, τ specify part of the agents' strategies. The speculative fraction $X_{i,t}$ of stocks $S_{i,t}$ which the agent would like to buy or sell (according to the sign) is

$$X_{i,t} = F_i[p_t, p_{t-1}, \ldots] \tag{3.7}$$

where $F_i[\cdot]$ is the strategy of agent i. In detail the function $F_i[\cdot]$ is a linear combination of moving time averages of derivatives of $\log p_t$.

Let us now interpret the two terms Eq. 3.6 is composed of. The first term on the right side of Eq. 3.6 mimics a chartist strategy: in fact the speculative action of the agents is the result of the observation and the analysis of the past trends. Agents primarily act as technical traders. As all trend-based strategies, their impact can be potentially destabilizing.

The second contribution of the agent's action $\Delta S_{i,t}$ on the market plays a role similar to rebalancer agents in Kim and Markowitz model. In fact this contribution tends to rebalance the portfolio composition in order that $S_{i,t}/M_{i,t} = \gamma_i/p_t$. The parameter τ sets the characteristic time that is required to perform the portfolio rebalancing. This term has a stabilizing effect because the higher will be the price, the lower will be the relative exposure on the risky stock (relative with respect to the cash amount $M_{i,t}$).

To sum up, once fixed the parametrization of the strategy and the function $F_i[\cdot]$, the strategy of the agent i is univocally determined by the set of parameters $\{\eta_1, \ldots, \eta_l, \tau, \gamma\}_i$. Thus, in this model agents are really heterogeneous with respect to the strategy, because the same price history leads to different actions differently from Kim and Markowitz model, where different decisions are the result of different signals and not of different strategies within the same class of agents.

Number of Agents

While the total number is kept constant, the number of active agents is not a static quantity differently from Kim and Markowitz model. In fact at each time step the poorest agent is removed from the market and replaced by a new agent endowed with new random strategies. It is sense the population of traders is dynamic.

However, the question of the non-stationarity of the number of active traders is not addressed in the model.

Price Formation

The feedback between prices and agents' actions is performed by a Walrasian mechanism of price update, that is the price response is proportional to the demand/supply imbalance.

At each time step t the previous clearing price is made public and each agent places an order according to Eq. 3.6. The excess demand ED_t is the sum of the demand $D_t = \sum_{i,\Delta S_{i,t}>0} \Delta S_{i,t}$ and the supply $O_t = \sum_{i,\Delta S_{i,t}<0} \Delta S_{i,t}$. The trading rules of the model are also provided with small buying transaction costs and small random fluctuations of the amount of agents' cash. The next price is fixed in order to be the price at which the excess demand is cleared and therefore

$$p_{t+1} = p_t \frac{\langle D_t \rangle}{\langle O_t \rangle} \tag{3.8}$$

where $\langle \cdot \rangle$ is an exponential moving average. This equation describes a multiplicative evolution of prices and this is why authors choose the logarithmic of the price as the price indicator.

Origin of Stylized Facts

We can argue that the competitive action which is present at each time step in agent's operations $\Delta S_{i,t}$ is the origin of the Stylized Facts in this model. In fact, if the model is suitably tuned, the two terms of Eq. 3.6 counterbalance each other giving rise to a market which shows fluctuations at all scales without the need of exogenous regulatory rules to stabilize the market behavior. This is probably the origin of the fat-tailed pdf of returns that decays with an exponent close to 2.

Moreover the model does not have simple arbitrage opportunities because the autocorrelation function of returns is substantially zero. In the end the return distribution shows a scaling exponent similar to the real one when the aggregation of returns is performed that is $F(r, \tau) = \tau^{0.62} g(r\tau^{-0.62})$ where τ is the time lag at which returns r are calculated. For the rank-size rule of agents' wealth the authors find a power law behavior with exponent very close to 1, as in real cases.

Discussion Realism/Tractability

Tractability is not a strong point of model. In fact the model requires a very fine tuning of strategies $F[\cdot]$ and of parameters in order to obtain a stable regime for prices, at least for the time scale investigated in the original paper ($10^3 - 10^4$) time steps.

In [33] it is argued that the model is asymptotically unstable that is the price converges to zero or diverges for $t \to \infty$). In this sense this model would not be really self-organized and would not be stable without external regulatory rules.

However, the time unit of the model is not explicitly calibrated thus the model could be substantially stable for times long enough to conclude that endogenous ingredients are sufficient to reproduce a self-regulating market.

With respect to realism, this model is one of the first ABM which is able to partially reproduce the main Stylized Facts in a quantitative way and has the merit to reach this goal introducing only a self-regulated market through the feedback between price evolution and agents' actions. To our knowledge it is perhaps the first model which underlines the non-trivial problem of a self-organized market. The model shows that a quasi-stable self-organized state can exist but does not answer the question of how this state is reached and how it is restored.

Self-Organized Criticality

This work is one of the first which deal with the self-organization of the market. The model produces a stable market given some particular classes of functions $F[p_t\, p_{t-1}, \ldots]$ and some very restricted areas of the phase space of its parameters. The authors argue that the simulated market is self-organized in the sense that external rules (i.e. exogenous forces) are not needed in order to obtain a realistic and stable market once the model is tuned.

3.3.5 Lux and Marchesi Model

Aim

The aim of the Lux-Marchesi model [34, 35] is to show that the scaling laws that are observed in financial markets can arise from agents' mutual interaction. In particular, Lux and Marchesi stress the clear difference between the statistical properties of the model input, i.e. the normal noise that makes the stock fundamental price evolve in time, and the output, that is the price dynamics produced by the agents' operations.

Agents and Strategies

The agents are divided in two categories: the **fundamentalists**, who believe in the existence of a fair price p_f for the traded stocks (and so, they sell stocks if $p(t) > p_f$ and they buy if $p(t) < p_f$), and **the chartists**, that are noise traders whose behavior is dictated by herding and historical prices. While the total number of agents N remains fixed throughout the simulation, the number of fundamentalists and chartists, n_f and n_c, are allowed to vary.

The effect of the two classes of agents on the price is very different: while fundamentalists have a stabilizing effect on the market, as their operations drive the price towards the fundamental one, chartists have a destabilizing effect and create bubbles and crashes. Furthermore, chartists agents are divided in two subcategories, optimists, who believe that the price will rise and hence always buy stocks, and pessimists, who believe that the price will decrease and so, on the contrary, always sell stocks.

At each time step each agent can change her category with a given transition probability. In this way, an internal dynamics within agents' classes is established.

The presence of the herding factor in the probability of transition among strategies introduces an intermittent dynamics between the populations and, as we will see in Sect. 3.3.5, is intimately connected to the emergence of the Stylized Facts.

The dynamics admits fixed points corresponding to zero agents in one of the categories. These cases are avoided by imposing suitable lower limits.

Price Formation

At each time step the price is increased or decreased of a tick (one cent) with a probability given by

$$\pi_{up} = \max[0, \beta(ED) + \mu] \qquad \pi_{down} = -\min[0, \beta(ED) + \mu]$$

respectively, where μ is a white noise that takes random traders into account and ED is the excess demand, that is, the difference between the aggregated operations on the market.

Stylized Facts and Their Origin

The herding terms in the transition probabilities between fundamentalists and chartists give rise to an alternation between the two metastable states in which one population is predominant. This process has been studied for the first time by Kirman [36] and then, in two asymmetric versions that are closer to the Lux-Marchesi dynamics, by Alfarano et al. [37] and Alfi et al. [38]. These simplified versions permit to solve analytically the contagion models proposed.

In the Lux-Marchesi model the most probable state is the one in which the majority of agents is fundamentalist. However, a stochastic fluctuation can trigger a bubble of chartism that can persist for a long time. During this period one finds a strong alternation between optimists and pessimists, because their rates are both proportional to $\frac{n_c}{N}$. Consequently, many buy and sell orders arrive in the market and the price goes quickly up and down accordingly, causing an increment of volatility. This causes an intermittency between regimes which have fluctuations with different amplitudes and, consequently, the emergence of fat tails and volatility clustering.

Realism and Tractability

Lux and Marchesi have included various reasonable aspects in their model: the importance of herding, the price evolution as a consequence of excess demand, the alternation of stabilizing and destabilizing factors in the market. In order to do this, they have built a complicated dynamics in which the roles played by the 13 parameters involved is not clear. This leads to a realistic dynamics, but reduces the tractability of the model. However, in [35] the properties of stationary states are analyzed. In these states the price is equal to the fundamental one and there are no net flows among the populations. The authors have studied the stability of these states and their relationships with the emergence of the Stylized Facts.

Number of Agents

In the Lux-Marchesi model the total number of agents N is fixed. On the contrary, the number of agents in each class can vary. Egenter et al. in [39] have shown that in the limit of large N the Lux-Marchesi model is no more able to reproduce the Stylized Facts. In this case all the agents become fundamentalists and so the price strictly follow the fundamental price. This mechanism can be easily understood if one considers a simplified asymmetric model for the population dynamics [38, 40, 41].

Self-Organized Criticality

The problem of self-organization is not addressed in this paper.

3.3.6 Giardina and Bouchaud Model

Aim

The model introduced in [42] is built to be, as explicitly declared by the authors, a *trait d'union* between the tractability of Minority Games and the complicated Sante Fe virtual stock market. However, the resulting model is still rather complicated and a fine tuning of the model parameters is required. In fact this model is characterized by a large amount of parameters but the Stylized Facts are only reproduced in a very small region of the parameter space. The novelty of this model is that for the first time the problem of the self-organization is, at least qualitatively, discussed and addressed.

Agents and Strategies

There are two tradable assets: a risky stock and a risk free bond with interest rate r.

As in Santa Fe artificial stock market agents inductively adapt their strategies and learn from their past performances. All agents share the same information, that is the price times series up to the present, and agents at each time step can revise their portfolio composition by selling, buying or not varying the number of owned shares.

The update of the portfolio is determined by the strategy with the best score at that time.

Number of Agents

The total number of agents is constant but agents can decide to be active or inactive on the market. Therefore the number of active agents changes along time but this number is bounded from above. In addition, in this model the number of active agents is substantially proportional to the trading volume. On this account the presence of the inactive strategy is crucial in order to reproduce the intermittent regime in which Stylized Facts are observed.

The authors of this model investigated a wide range of N up to 10,000 but they do not observe a significant dependence of the model features on N, differently from Lux and Marchesi model.

Origin of Stylized Facts

The model is characterized by three regimes which depend mainly on two parameters: the ratio g/λ which fixes the market response to an imbalance and the polarization P of the chartist strategies.

A *rational* regime is observed for large negative polarization (P close to -1) because the chartist strategies are almost all mean reverting and the stock price follows the fundamental one.

An oscillating regime is found for $g/\lambda < 0.4$ and positive polarization, where periodic bubbles and successive crashes take place. This regime is characterized by large arbitrage opportunities and inefficiency.

The most interesting regime which is obtained for $g/\lambda \geq 0.4$ and $P \geq -|P_0|$ ($|P_0| \sim 0.05$) reproduces most of the market Stylized Facts and is almost efficient.

We now focus our attention on this intermittent regime which exhibits realistic features. The price time series shows volatility clustering (only for large values of g/λ) and fat tails (the exponent of the tail distribution of returns is around 3.5).

The investigation of the scaling properties of returns with respect to the time lag Δt on which the return is performed reveal a Ornstein-Uhlenbeck-like behavior, i.e. the price is mean-reverting on long time horizon and consequently the price fluctuations become constant and independent of Δt for large values of the time lag. This mean-reverting tendency is due to the finite wealth that cannot indefinitely fuel a divergence from the fundamental price. In fact the same behavior is observed when all the agents' strategies are random and therefore this effect is only due to the constraint represented by the finite capital. Such a convergence to a constant value is not observed in real price time series.

The origin of the volatility clustering can be traced back to the fact that volume and volatility are correlated only in the high activity region while they result uncorrelated in low activity time. This non linear relationship produces a non trivial behavior for the volatility (see [43] for further details).

Discussion Realism/Tractability

The model matches most of the Stylized Facts but, as Lux and Marchesi model, its building blocks are far from being the very minimal ones. Despite it can be approximately described by only two quantities P and g/λ and allows for some analytic approaches in some regimes, the model has never been used for a systematic investigation of the market properties.

Self-Organized Criticality

To our knowledge this is the first work that really addresses the question of the dynamics of the self-organization of the markets towards the intermittent state of the Stylized Facts. Unfortunately the authors does not give an answer to this problem because they do not investigate the interesting argument proposed in [42] in the framework of their model. In fact they observe that in principle for small values of the ratio g/λ the model produces very long bubbles and a reasonable behavior for investors in such a scenario would consist in increasing the invested fraction g (the model rules do not allow for such a dependence, g is fixed along the simulation) and therefore g/λ will spontaneously increase. We recall that the intermittent regime is observed for $g/\lambda \geq 0.4$. Meanwhile small values of g/λ also produces small rate of executions and therefore the market will tend to be more liquid that is λ will decrease. Both the effects contribute to enhance the ratio g/λ and the market enter into the intermittent regime. Once this quasi efficient regime with burst of activity is reached the ratio should become almost constant because the quasi-efficiency and the burst of volatility does not fuel anymore the growth of g/λ.

In the end it is interesting that the intermittent regime is obtained when the chartists are not mean reverting on average. This suggests that the destabilizing effect of trend following strategies is crucial to understand and reproduce the intermittency of real markets. A similar conclusion is drawn in the model proposed in [38, 40, 41, 44].

3.4 Summary and Perspectives

Broadly speaking one can define two variables: realism and tractability and in Fig. 3.2 we have given a pictorial representation of the models in this respect.

Which Model is Better?

This question has no clear answer because the Stylized Facts, which represent the experimental framework are relatively limited and not too difficult to reproduce in ABM. Therefore a first conclusion is that progress in this area is directly related to the possibility of increasing the number and the quality of the Stylized Facts. This would provide additional tests of the models and permit a quality assessment of them.

How Many Parameters?

This question relates to the fundamental dilemma of realism versus simplicity and has no simple answer. Clearly in terms of metaphors the minimum number of parameters is optimal to grasp a concept in a simple way. This has been the philosophy of the fundamental models for critical phenomena in Statistical Physics and Complex Science [8]. On the other hand, even a simplified description of real markets may require more elements specific to this field which may not have a counterpart in Statistical Physics.

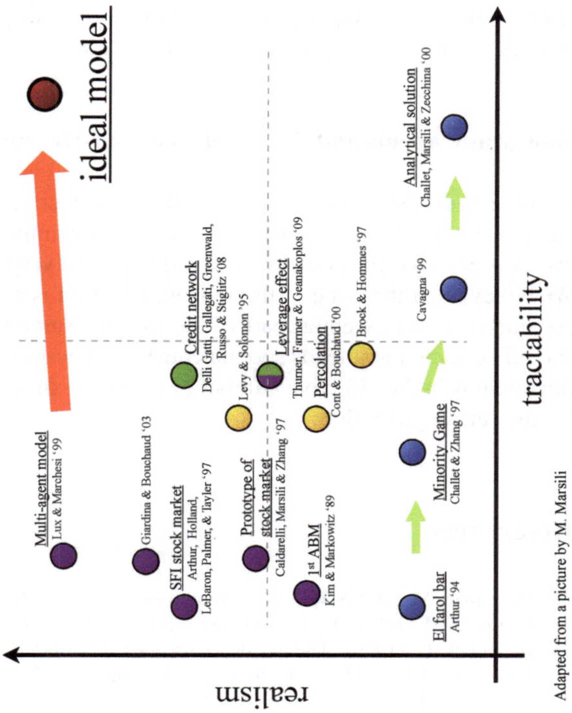

Fig. 3.2 In this picture we report several ABM organized along the two axis of realism and tractability. The picture was inspired by a talk by Matteo Marsili and is intended to show the rather different perspectives of the various models together with the path we have tried to follow in constructing our approach. The Ideal Model is still far and the available data (Stylized Facts) are too scarce to permit a quality assessment of the various models

Universality

The Stylized Facts observed, characterized essentially by fat tails and volatility clustering resemble the power law behavior of Critical Phenomena. This leads to the fundamental issue of Universality. Clean power laws in fact are observed as asymptotic properties and, in this limit, they may be universal in the sense that do not depend on the details of the model but only on its basic principles. Many physicists like the idea that the same may happen in Economics and that the observed deviations from universality in the data depend only on the poor quality of these.

In our opinion we are in an intermediate situation. Certain properties of the various markets appear qualitatively similar but their quantitative properties are really different. This also touches the question whether the power laws are approximate or genuine and clearly we point to the first hypothesis.

The basic reason for this point of view is that in most of the models considered the Stylized Facts do not correspond to a genuine asymptotic behavior and they are can be obtained for a very specific number of agents.

This situation of similarity but not strict universality has advantages and disadvantages. On one hand one looses the clean properties of universal exponents.

On the other hand one acquires a sort of biodiversity which may lead to a specific characterization of the properties of each market.

How Many Agents and the Problem of Self-Organization

In some agent-based models the number of effective agents is fixed at some finite value while in others it can vary in a more or less important way. In all models, however, the Stylized Facts are reproduced only in a very limited region of parameters. We believe that the issue of the self-organization is a crucial point which should be considered as an essential part of the model. Namely the dynamics of the system should be such that it leads spontaneously to the realistic region. We believe that the fluctuations in the effective number of agents or strategies are an important concept for the self-organization.

References

1. Buchanan, M. (2008). Why economic theory is out of whack. *New Scientist, 2665*, 32.
2. Kreps, D. M. (1990). *A course in microeconomic theory*. Princeton: Princeton University Press.
3. Bachelier, L. (1900). Théorie de la spéculation. *Annales Scientifiques de l'Ecole Normale Supérieure* III-17, 21. (Ph.D. thesis is mathematics).
4. Feller, W. (1968). *An introduction to probabbility and its applications* (Vol. 1). New York: Wiley.
5. Black, F., & Scholes, M. (1973). The pricing of options and corporate liabilities. *Journal of Political Economy, 81*, 637.
6. Hull, J. C. (1997). *Options features and other derivatives*. Upper Saddle River NJ: Prentice-Hall.
7. Cutler, D. M., Poterba, J. M., & Summers, L. H. (1989). What moves stock prices? *The Journal of Portfolio Management, 15*, 4.
8. Pietronero, L. (2008). Complexity ideas from condensed matter and statistical physics. *Europhysics News, 39*, 26.
9. Bouchaud, J.-P. (2008). Economics needs a scientific revolution. *Nature, 455*, 1181.
10. Farmer, J. D., & Foley, D. (2009). The economy needs agent-based modelling. *Nature, 460*, 685.
11. Arrow, J. K., & Debreu, G. (1954). Existence of an equilibrium for a competitive economy. *Econometrica, 22*, 265.
12. Debreu, G. (1959). *Theory of value*. New York: Wiley.
13. Arrow, J. K. (1953). Le role des valeurs boursieres pour la repartition la meilleure des risques. *Econometrie, 11*, 41.
14. Arthur, W., Holland, J. H., LeBaron, B., Palmer, R., & Tyler, P. (1996). Asset pricing under endogenous expectation in an artificial stock market. Retrieved from http://www2.econ.iastate.edu/tesfatsi/ahlpt96.pdf.
15. Chakraborti, A., Toke, I. M., Patriarca, M., & Abergel, F. (2011) Econophysics: Empirical facts and agent-based models. *Quantitative Finance* (in press). Retrieved from http://arxiv.org/abs/arXiv:0909.1974v2.
16. Samanidou, E., Zschischang, E., Stauffer, D., & Lux, T. (2007). Microscopic models of financial markets. *Reports on Progress in Physics, 70*, 409–450.

17. Hommes, C. H. (2005). Heterogeneous agent models in economics and finance. Tinbergen Institute Discussion Papers 05–056/1, Tinbergen Institute. Retrieved from: http://ideas.repec.org/p/dgr/uvatin/20050056.html.
18. Report of the presidential task force on market mechanisms. (1988). *Tech.* The Brady Commission, Washington DC : Rep. US Government Commission Printing Office.
19. Schwert, G. W. (1990). Stock volatility and the crash of 87. *Review of Financial Studies*, *3*, 77.
20. Kim, G., & Markowitz, H. M. (1989). Investment rules, margin and market volatility. *Journal of Portfolio Management*, *16*, 45.
21. Brennan, M. J., & Schwartz, E. S. (1989). Portfolio insurance and financial market equilibrium. *Journal of Business*, *62*, 455.
22. Palmer, R., Arthur, W., Holland, J. H., LeBaron, B., & Tyler, P. (1994). Artificial economic life: a simple model of a stockmarket. *Physica D*, *75*, 264.
23. Roehner, B. M. (2002). *Patterns of speculation*. Cambridge: Cambridge Univeristy Press.
24. Challet, D., Marsili, M., & Zhang, Y.-C. (2005). *Minority games: Interacting agents in financial markets*. New York, NY, USA: Oxford University Press, Inc.
25. De Martino, A. (2006). Statistical mechanics of socio-economic systems with heterogeneous agents. *Journal of Physics A: Mathematical and General*, *39*, R465.
26. Arthur, B. W. (1994). Inductive reasoning and bounded rationality: The el farol problem. *American Economic Review*, *84*, 406.
27. Challet, D., & Zhang, Y. (1997). Emergence of cooperation and organization in an evolutionary game. *Physica A*, *246*, 407.
28. Challet, D., Marsili, M., & Zhang, Y.-C. (2001). Stylized facts of financial markets and market crashes in minority games. *Physica A: Statistical Mechanics and its Applications*, *294*(3), 514–524.
29. Challet, D., & Marsili, M. (2003). Criticality and market efficiency in a simple realistic model of the stock market. *Physical Review E*, *68*, 036132.
30. Challet, D., Marsili, M., & Zecchina, R. (2000). Statistical mechanics of systems with heterogeneous agents: Minority games. *Physical Review Letters*, *84*, 1824.
31. Mezard, M., Parisi, G., & Virasoro, M. (1986). Spin glass theory and beyond: an introduction to the replica method and its applications (World Scientific Lecture Notes in Physics).
32. Caldarelli, G., Marsili, M., & Zhang, Y.-C. (1997). A prototype model of stock exchange. *Europhysics Letters*, *40*, 479.
33. Golke, S., Preis, T., Paul, W., & Schneider, J. J. Comment on a prototype model of stock exchange by G. Caldarelli, M. Marsili, and Y.-C. Zhang. Retrieved from http://www.staff.uni-mainz.de/schneidj/papers/papergolke.pdf.
34. Lux, T., & Marchesi, M. (1999). Scaling and criticality in a stochastic multi-agent model of a financial market. *Nature*, *397*, 498.
35. Lux, T., & Marchesi, M. (2000). Volatility clustering in financial markets: A micro-simulation of interacting agents. *International Journal of Theoretical and Applied Finance*, *3*, 675.
36. Kirman, A. (1993). Ants, rationality and recruitment. *The Quarterly Journal of Economics*, *108*, 137.
37. Alfarano, S., Lux, T., & Wagner, F. (2005). Estimation of agent-based models: the case of an asymmetric herding model. *Computational Economics*, *26*, 19.
38. Alfi, V., Cristelli, M., Pietronero, L., & Zaccaria, A. (2009). Minimal agent based model for financial markets I: origin and self-organization of stylized facts. *European Physical Journal B*, *67*, 385. http://arxiv.org/abs/0808.3562.
39. Egenter, E., Lux, T., & Stauffer, D. (1999). Finite-size effects in monte carlo simulations of two stock market models. *Physica A*, *268*, 250.
40. Alfi, V., Pietronero, L., & Zaccaria, A. (2009). Self-organization for the stylized facts and finite-size effects in a financial-market model. *EPL*, *86*, 58003.
41. Alfi, V., Cristelli, M., Pietronero, L., & Zaccaria, A. (2009). Minimal agent based model for financial markets II: statistical properties of the linear and multiplicative dynamics. *European Physical Journal B*, *67*, 399. http://arxiv.org/abs/0808.3565.

42. Giardina, I., & Bouchaud, J.-P. (2003). Bubbles, crashes and intemittency in agent based market. *European Physical Journal B*, *31*, 421.
43. Bouchaud, J.-P., Giardina, I., & Mézard, M. (2001). On a universal mechanism for long ranged volatility correlations. *Quantitative Finance*, *1*, 212.
44. Alfi, V., Cristelli, M., Pietronero, L., & Zaccaria, A. (2009). Mechanisms of self-organization and finite size effects in a minimal agent based model. *Journal of Statistical Mechanics*, P03016. doi:10.1088/1742-5468/2009/03/P03016

Chapter 4
A Minimal Agent-Based Model and Self-Organization of Financial Markets

In the past years there has been a large interest in the development of Agent Based Models (ABM) aimed at reproducing and understanding the Stylized Facts observed in the financial time series. The simplest model of price time series is the Random Walk introduced by Louis Bachelier in 1900 [1]. The availability of large amounts of data has revealed a variety of systematic deviations from the Random Walk which are the Stylized Facts and are relatively common to all markets (see Chap. 2). The properties of the Stylized Facts resemble Critical Phenomena and Complex Systems with possible relations to various problems in Statistical Physics. Many questions are however open and also how far one can go with these similarities represents a highly debated problem. Beyond the Stylized Facts reported in Chap. 2 a point which is usually neglected is the fact that the Stylized Facts usually correspond to a particular situation of the market. If the market is pushed outside such a situation, it will evolve to restore it spontaneously. The question is why and how all markets seem to self-organize in this special state. The answer to this question represents a fundamental point in understanding the origin of the Stylized Facts.

In this chapter we discuss a minimal ABM which includes the following elements:

- **fundamentalists:** these agents have as reference a fundamental price p_f derived from standard economic analysis of the value of the stock. Their strategy is to trade on the fluctuations from this reference value and bet on the fact that the price will finally converge towards the reference value. These traders are mostly institutional and their time scale is relatively long. Their effect is a tendency to stabilize the price around the reference value.
- **chartists:** they consider only the price time series and tend to follow trends in the positive or negative direction. In this respect they induce a destabilizing tendency in the market. These traders have usually a time horizon shorter than the fundamentalists and they are responsible for the large price fluctuations corresponding to bubbles or crashes.
- **herding effect:** this is the tendency to follow the strategy of the other traders. It should be noted that traders can change their strategy from fundamentalist to chartist and vice-versa depending on various elements.

M. Cristelli, *Complexity in Financial Markets*,
Springer Theses, DOI: 10.1007/978-3-319-00723-6_4,

- **price behavior:** each trader looks at the price from her perspective and derives a signal from the price behavior. This signal will be crucial in deciding her trading strategy.

We believe that these four elements are the essential irreducible ones that an ABM should possess. Of course real markets are extremely more complex but the study of these elements represents a basis on which one might eventually add more realistic features.

In this perspective we construct a model which is the simplest possible containing these four elements. This permits to obtain a detailed understanding of the origin and nature of the Stylized Facts and also to discuss their Self-Organization.

4.1 Definition of a Minimal ABM with Fundamentalists and Chartists

The model is constructed to be as simple as possible but still able to reproduce the Stylized Facts of financial time series. This simplicity is very important because it permits to derive a microscopic explanation of the Stylized Facts in term of the model parameters. This model contains the main ingredients of the well known Lux-Marchesi (LM) model [2, 3] but differently from this one it has the great advantage to be a very minimal model which permits analytical approaches in some simple cases and above all it is a "workable model" in the sense that it can be easily modified to introduce variants and elements of realism. In fact, even though the LM class of models include the elements we consider fundamental for an agent based model to have, its complicate architecture prevents to consider systematic developments of the model.

In our model we consider a population of N interacting agents which are divided into two main categories: fundamentalists and chartists. The fundamentalist agents tend to stabilize the market driving the price towards a sort of reference value which we call the fundamental price (p_f). This kind of agents can be identified with long-term traders and institutional traders [4]. Chartists instead are short-term traders who look for detecting a local trend in the price fluctuations. This kind of agents tries to gain from detecting a trend and so they are responsible for the formation of market bubbles and crashes which destabilize the market. In the LM model the destabilizing effect made by chartists is implemented dividing the chartists agents into two subcategories: optimists which always buy and pessimists which always sell. In our model we can overcome this complication by simplifying the chartists' behavior. In fact, the description of chartists is in term of the recently introduced potential method [5–8]. Chartist agents detect a trend by looking at the distance between the price and its smoothed profile which in our case is the moving average computed on the previous M time steps. Chartists try to follow the trend and bet that the price will moreover move away from the actual price. In this way they create a local bubble which destabilizes the market. The stochastic equation for the price which describes the chartists' behavior can be written in terms of a Random Walk with a force whose center is the moving average of the price:

$$p_{t+1} = p_t + \frac{b}{M-1} F(p_t - p_{M_t}) + \sigma \xi_t \tag{4.1}$$

$$p_{M_t} = \frac{1}{M} \sum_{i=0}^{M-1} p_{t-i} \tag{4.2}$$

where b is a parameter which gives the strength of the force, p_M is the moving average performed on the previous M time steps, ξ is a white noise and σ is the amplitude of this noise. The factor $M - 1$ in the denominator of the force term makes the potential independent of the particular choice of the moving average length [5]. This equation has been used in previous papers to analyze real data [5, 7, 8]. The conclusion is that is indeed possible to observe this kind of forces also in real stock-prices. For our model we decided to adopt a simple linear expression for the force ($F = p_t - p_{M_t}$). The reason supporting such an assumption (i.e. a linear force) can be traced back to a series of *experimental* works [6, 9–11, 7] which measured whether or not there exists a repulsive/attractive effect due to the moving average. One of the main result of these works is that real time series (in detail yen/US dollar and a selection of stocks traded at NYSE) exhibit this kind of hidden forces which can be explained in terms of deviations from the moving average. In addition these papers also investigate how the potentials of attraction/repulsion with respect to the moving average are shaped. In practice the analysis of the potential is preferred to the direct analysis of the force for technical reasons (in fact the potential is the result of an integral therefore it is generally a more smooth function than the force). It is found that the shape of the potential is approximately parabolic or, in general, a parabolic behavior (i. e. linear force) appears to be a good approximation.

By integrating the force it is also possible to obtain the effective potential which describes the chartists' behavior. For the case of a simple linear force the potential is quadratic. In Fig. 4.1 we show the stochastic process described in Eqs. 4.1 and 4.2 with a linear force and the corresponding potential. A random walk is also plotted for comparison.

Depending on the sign of the parameter b one can obtain both attractive and repulsive potential. Here we are considering only the repulsive potential which describes the trend-follower behavior of chartists, because the stabilizing attractive case is already carried out by the fundamentalist agents. In fact, fundamentalists try to stabilize the market driving the price towards the fundamental price p_f which in our model is constant in time. The stochastic equation which describes the fundamentalists' behavior can be written in terms of a random walk with a further term which is responsible for the stabilizing action of fundamentalists:

$$p_{t+1} = p_t + \gamma(p_f - p_t) + \sigma \xi_t \tag{4.3}$$

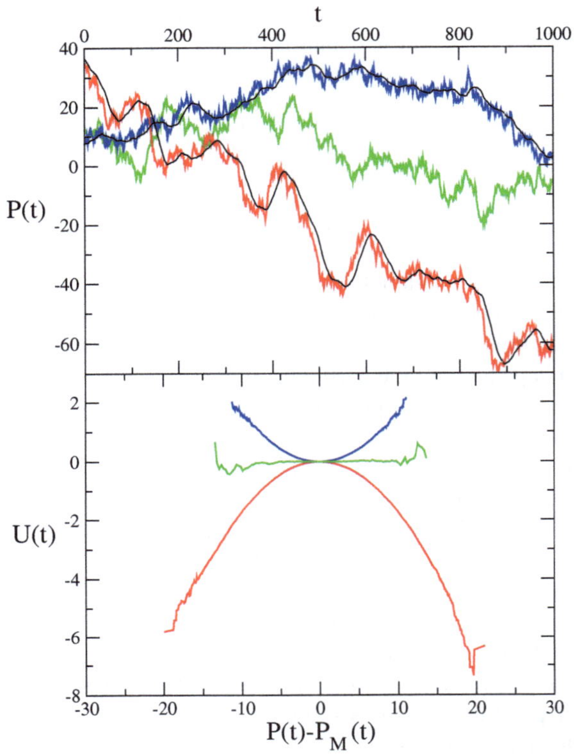

Fig. 4.1 The upper plot shows the price fluctuations for the process described by Eqs. 4.1 and 4.2 with a linear force F for different values of the force strength parameter b. The values of the parameter b are $b = -1$ (*blue line*), $b = 0$ (*green line*) and $b = 1$ (*red line*) which correspond respectively to attractive, flat and repulsive potentials as shown in the lower plot. In our model we are considering only the repulsive potential which describes the behavior of the trend-follower chartists

where γ is the strength of the fundamentalists' action. For the moment the total number of agents, N, is kept fixed and there are N_f fundamentalists and N_c chartists but agents can decide to change their mind during the simulation by switching their strategy from fundamentalist to chartist and vice-versa. The probability to change strategy is based on two terms. The first is an herding term: agents tend to imitate other people behavior proportionally to the relative number of agents in the arrival class. The second is a term, characterized by the parameter K, which leaves the agents the possibility to change strategy on the basis of considerations on the price behavior, independently on what the other agents are doing. The price signal which appears in the transition probabilities is proportional to $|p_t - p_f|$ for the probability to become fundamentalist and $|p_t - p_{M_t}|$ for that to become chartists. The mathematical expression for the transition probabilities is:

$$P_{cf} \propto \left(K + \frac{N_f}{N}\right) \exp\left(\gamma |p_t - p_f|\right) \tag{4.4}$$

$$P_{fc} \propto \left(K + \frac{N_c}{N}\right) \exp\left(\frac{b}{M-1} |p_t - p_{M_t}|\right) \tag{4.5}$$

For a realistic representation of the market the fundamentalists should dominate for very long times with intermittent appearance of bubbles or crashes due to the chartists. This is due to the fact that fundamentalists are usually big institutional traders which have on average a major weight in the market. In order to properly reproduce this effect we can introduce an asymmetry in the herding dynamics and, neglecting the price modulation, Eqs. 4.4 and 4.5 can be written in the following way:

$$P_{cf} \propto \left(K + \frac{N_f}{N}\right)(1+\delta) \tag{4.6}$$

$$P_{fc} \propto \left(K + \frac{N_c}{N}\right)(1-\delta) \tag{4.7}$$

where the positive parameter δ define the asymmetry of the model. In the case $\delta = 0$ the model reduces to the symmetric ants model by Kirman [12]. In Sect. 4.2 we will analyze in detail this asymmetric model and the corresponding equilibrium distribution for the population of chartists and fundamentalists. For the price formation we are going to use a simple linearized form of the Walras' price adjustment mechanism [13]:

$$p_{t+1} - p_t = \Delta p \propto ED_t \tag{4.8}$$

where ED is the excess demand of the market at time t. This linear interpretation, valid for small price increments, ($\Delta p \ll 1$) is a technical simplification from the more realistic multiplicative dynamics and it is very useful for an analytical treatment. With respect to the dynamics of the ABM the multiplicative case will turn out to be less stable in terms of parameters while the the linear case is mathematically less problematic.

In our model the excess demand ED is simply proportional to the price signal of chartists and fundamentalists:

$$ED = ED_f + ED_c = \frac{N_f}{N}\gamma(p_f - p_t) + \frac{N_c}{N}\frac{b}{M-1}(p_t - p_{M_t}) \tag{4.9}$$

then adding a noise term we have the complete equation for the price formation:

$$p_{t+1} - p_t = ED + \sigma\xi_t \tag{4.10}$$

In this case we have that the volume of the agents' actions is proportional to the signal they perceive. This situation is much simpler than the corresponding price formation for the LM model where prices can only vary of a fixed amount (tick) and the connection with the excess demand is implemented in a probabilistic way.

4.2 Herding Dynamics

4.2.1 Symmetric Case

In his seminal paper Kirman [12] proposed a simple dynamical model to explain a peculiar behavior observed in ant colonies. Having two identical food sources, ants prefer alternatively only one of these, and (almost) periodically switch from one source to the other. In Kirman's model this effect is caused by an *herding dynamics*, in which the evolution is stochastic and based on the meetings of the ants. In particular an ant can recruit a companion, and bring it to its preferred source, with a certain probability.

This model can be used to take into account the herding dynamics of a population of N agents (with N fixed), and it is formalized as follows.

Let us suppose the existence of two microscopic states: an agent at time t can be either a *fundamentalist* or a *chartist*. Defining $N_c = N_{ct}$ as the number of chartists at time t (and analogously $N_f = N_{f_t} = N - N_{ct}$ as the the number of fundamentalists), the quantity

$$x = \frac{N_c}{N} \tag{4.11}$$

varying between 0 and $+1$ describes the macroscopic state of the system.

A possibility to change opinion is given to each agent at any given time step if she meets an agent with opposite views who succeeds in recruiting her. Under the hypothesis that the transition probabilities are small enough, in a single time step there will be no more than one change of opinion. In this case we can write the transition rates as

$$P_{cf} = \beta \frac{N_f}{N} \qquad P_{fc} = \beta \frac{N_c}{N} \tag{4.12}$$

where P_{cf} is the probability than *one* agent passes from the chartist group to the fundamentalist one, and β regulates the speed of the process.

If we adopt the rates of Eq. 4.12 the dynamics will admit two absorbing states ($N_c = 0$ and $N_c = N$). In [3, 2] a lower limit of 4 for each class of agents is assumed, while Alfarano et al. [14], following Kirman [12], suggested a Poissonian term (that is, independent from the number of agents) to avoid these fixed points. This new term can be seen as a spontaneous tendency to change one's mind, independently on the others. This corresponds essentially on the term K we have discussed in relation of Eqs. 4.4–4.7. The simplest model is therefore given by the following transition probabilities,

$$P_{cf} = \beta \left(K + \frac{N_f}{N} \right) \qquad P_{fc} = \beta \left(K + \frac{N_c}{N} \right). \tag{4.13}$$

Simulating this process one can observe two different behaviors depending on the choice of the parameters, and in particular on $\epsilon \equiv KN$. It can be shown [12] that if $\epsilon < 1$ the system spends most of the time in the two *metastable states* $x \approx 0$ and

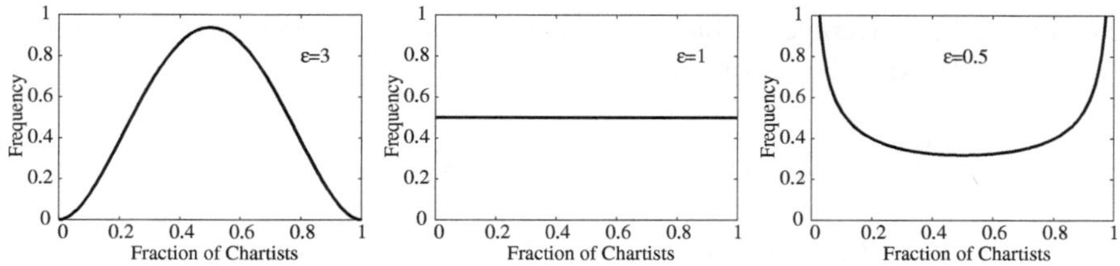

Fig. 4.2 The functional form of the distribution Eq. 4.14 changes with $\epsilon = KN$. We show the cases $\epsilon = 3.0$; 1.0 and 0.5. Only for $\epsilon < 1$ the system shows an alternation between two metastable states

$x \approx 1$, sometimes switching from one to the other. In the case $\epsilon \geq 1$, that is, if the self-conversion term is large enough, the systems fluctuates around $x = 0.5$.

Kirman explicitly derived the functional form of the equilibrium function for x,

$$P_s(x) \sim x^{\epsilon-1}(1-x)^{\epsilon-1}. \tag{4.14}$$

In Fig. 4.2 the three cases $\epsilon > 1$, $\epsilon = 1$ and $\epsilon < 1$ are shown. In the first case the distribution has a single peek for $x = 0.5$. By decreasing ϵ the distribution gets smoother and smoother, until it becomes uniform for $\epsilon = 1$. If one further decreases ϵ, the distribution develops two peeks and becomes bimodal.

4.2.2 Asymmetric Case

It is possible to generalize the symmetric herding model described above to the asymmetric case, in order to take account of the fact that institutional traders, that have more impact on the market than individuals, usually adopt long-term strategies [4], as fundamentalists do. Here we propose a further generalization of the model proposed by Alfarano et al. [15], which is more convenient for our purposes.

Our asymmetric population dynamics is given by:

$$p_{f\to c} = \beta(1-\delta)\left[K_1 + \frac{N_c}{N}\right] \tag{4.15}$$

$$p_{c\to f} = \beta(1+\delta)\left[K_2 + \frac{N_f}{N}\right] \tag{4.16}$$

where $\delta > 0$ regulates the asymmetry between the two metastable states (the model described in [15] can be recovered in the case $\delta = 0$).

We are going to see that the asymmetry introduced by the term δ is more realistic than the one due to K_1 and K_2 with respect to the objective of having the market dominated (on average) by fundamentalists at very large times.

It is possible to find the Fokker-Planck equation associated to the process given by Eq. 4.16. Following Alfarano et al. [15], we consider the Master Equation in the case in which, on average, we have no more than a change of opinion in a single time step (this approximation is valid if $\beta \ll 1$):

$$\frac{\Delta P_{N_c}(t)}{\Delta t} = P_{N_c+1}\pi(N_c+1 \to N_c) - P_{N_c}\pi(N_c \to N_c - 1) \tag{4.17}$$

$$- P_{N_c}\pi(N_c \to N_c + 1) + P_{N_c-1}\pi(N_c - 1 \to N_c). \tag{4.18}$$

where $P_{N_c}(t)$ is the probability to have N_c chartists at time t and π is the rate of the transition in the brackets. We can consider the continuous limit of Eq. 4.18 and recover the Fokker-Planck equation

$$\frac{\partial p(x,t)}{\partial t} = -\frac{\partial}{\partial x}A(x)p(x,t) + \frac{1}{2}\frac{\partial^2}{\partial x^2}D(x)p(x,t) \tag{4.19}$$

with drift and diffusion function given respectively by

$$A(x) = \beta\left[-2\delta x(1-x) + (1-\delta)K_1(1-x) - (1+\delta)K_2 x - \delta\frac{1}{N^2}\right] \tag{4.20}$$

and

$$D(x) = \beta\left[\frac{2}{N}x(1-x) + \frac{(1-\delta)}{N}K_1(1-x) + \frac{(1+\delta)}{N}K_2 x - 2\frac{\delta}{N^2}\left(x - \frac{1}{2}\right)\right]. \tag{4.21}$$

Given these functions, the asymptotic stable distribution for the fraction x of chartists is given by the formula

$$p_s(x,N) = \frac{C}{D(x)}\exp\left[\int^x \frac{2A(y)}{D(y)}dy\right] \tag{4.22}$$

where C is a normalization constant (this expression can be obtained setting the temporal derivative of $p_s(x,N)$ equal to zero and integrating twice the Fokker-Planck equation, see for example [16]).

The explicit expression of the equilibrium distribution can be derived by appropriate analytical computer codes and it is rather complex. Nevertheless a simple approximation will allow us to derive the distribution $\tilde{p}_s(x,N)$ which can be easily calculated and it illustrates clearly the role played by the asymmetry parameter δ.

The idea is to disregard the terms of order $1/N^2$ in the drift and in the diffusion functions. This approximation is valid for $N \gg 1$ and, since we will deal with at least ~50 agents, we can expect that it is rather appropriate to our case.

Moreover, we are interested only in the case $K_1 N = K_2 N < 1$ in order to avoid the unimodality of the asymptotic distribution. This can be obtained setting

$K_1 = K_2 = r/N$, with $r < 1$. The reasons for adopting a parametrization in which K is inversely proportional to N are:

(i) qualitatively one may expect that the probability to neglect the herding behavior decreases when N increases.
(ii) in the following we will consider the properties of the model as a function of N and we believe it is realistic that the system always stays in the bimodal case. This requires that $\epsilon = KN$ should not exceed the value one even for large values of N.

With this choice and using the approximation described above, the drift and the diffusion functions become respectively

$$\tilde{A}(x) = \beta\Big[-2\delta x(1-x) + \frac{r}{N}\Big(1-\delta-2x\Big)\Big] \tag{4.23}$$

and

$$\tilde{D}(x) = \beta\Big[\frac{2}{N}x(1-x)\Big] \tag{4.24}$$

Now we can easily solve the integral in Eq. 4.22 and derive

$$\tilde{p}_s(x, N) \propto x^{r(1-\delta)-1}(1-x)^{r(1+\delta)-1}\exp(-2\delta N x). \tag{4.25}$$

The form of Eq. 4.25 clearly shows that, by increasing N, the values of x near to 1 are exponentially suppressed. In other words, if N is small the equilibrium distribution will be bimodal, because $\delta \ll 1$ and $r < 1$ (compare the Eqs. 4.25 and 4.14), while by increasing N the asymmetry, regulated by the parameter δ, becomes more and more important.

We have also simulated the process described by the Eq. 4.16 and we have compared the distributions obtained by integrating the Eq. 4.22 with the complete drift and diffusion functions. As it is clear from Fig. 4.3, the agreement between theory and simulation is very good.

4.2.3 Finite Size Effects

Alfarano et al. [14] computed the time that on average must be waited to see the switch from one metastable state to the other in the symmetric case:

$$T_0 = \frac{N}{\beta}\frac{\pi}{(1-2\epsilon)}\frac{\cos(\pi\epsilon)}{\sin(\pi\epsilon)}. \tag{4.26}$$

The presence of this characteristic temporal scale implies that this simple model would give exponential correlation functions for volatility clustering. However

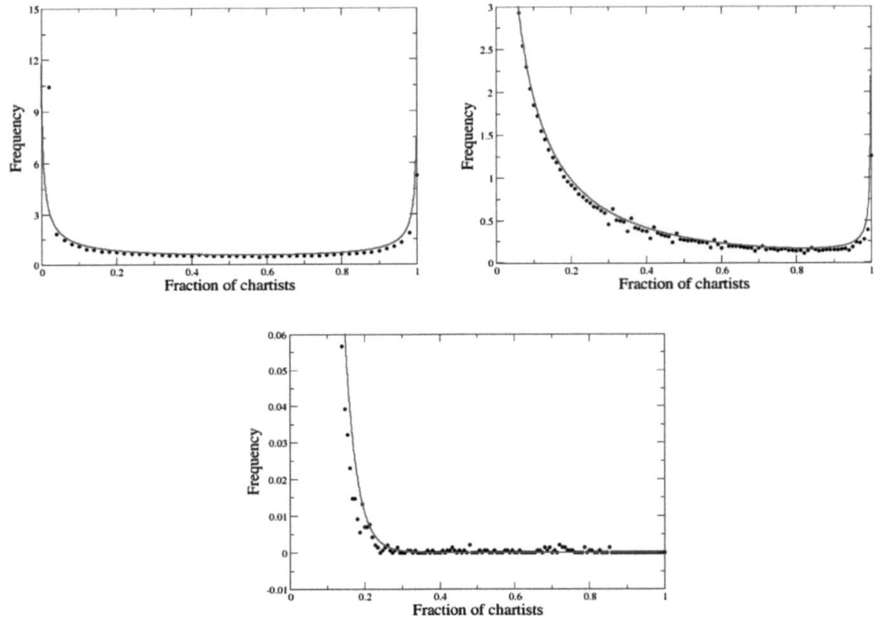

Fig. 4.3 Comparison between analytical approximation (*red line*) and simulation (*black circles*) for the asymmetric model described by the transition rates given by Eq. 4.16. From top to bottom the three cases $N = 50$, $N = 500$ and $N = 5000$ are shown, the other relevant parameter δ being fixed and equal to 0.003. The variation of the vertical scale underlines the progressive compression of the distribution towards the state in which fundamentalists are dominant

different time scales in agents' strategies can lead to a superposition of different characteristic times and therefore to long-range relaxation.

Examining the functional form of T_0 one can see that for $\epsilon \rightarrow 0$ the mean first passage time diverges, because of the emergence of the two absorbing states. Moreover, keeping ϵ constant and increasing N the time spent in one of the metastable states gets longer.

We have simulated the process given by Eq. 4.12 using different values of N keeping ϵ fixed and equal to 0.5. For the velocity parameter β we choose the value 0.02. As expected, the system oscillates between the two metastable states. As shown in Fig. 4.4, the only difference is the frequency of the passages from one state to the other. For small values of N ($N = 50$) the rate to change strategy is very high and this leads to an unrealistic situation in which fluctuations are too fast. On the other hand, for large values of N ($N = 5000$) the system gets essentially locked in one of the two states and the fluctuations become frozen. Only for an intermediate values of N ($N = 500$) the system shows an intermittent behavior which resembles experimental observations and will lead to the Stylized Facts. This finite size effect was first noted in [17] in relation to the LM model. In conclusion, this simple model explains why,

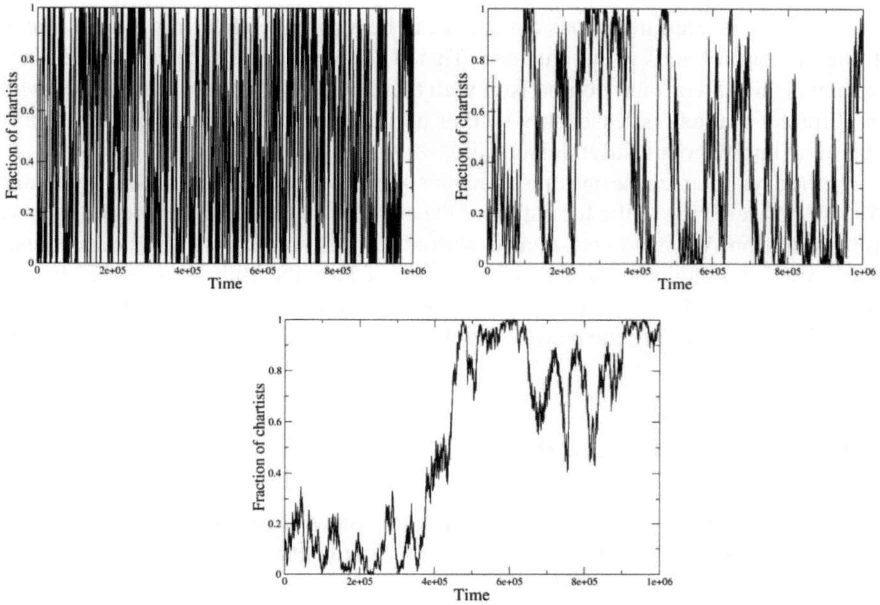

Fig. 4.4 Fraction of chartists x for the dynamics given by Eq. 4.12. We have considered different values of N but the same temporal scale for the horizontal axis, $T = 10^6$ time steps. From top to bottom we have $N = 50$, 500 and 5000 respectively. One can see that the cases $N = 50$ and $N = 5000$ are unrealistic, corresponding to too many or too few fluctuations. Only the case $N = 500$ shows the right intermittency which leads to stylized facts

in the limit $N \to \infty$, models like the LM one loose their properties to generate the Stylized Facts.

This result is based on very general properties and it is easy to expect that it will be a general one for the entire class of model we are considering. This implies that in this kind of models, given a fixed temporal scale, the system will be able to generate the "right" amount of fluctuations only within a specific range of finite values of N. This is a very important observation from both a conceptual and a practical point of view. In fact in this class of models the quasi-critical state linked to the Stylized Facts, corresponds to a finite size effect (with respect to N and to t) in the sense of Statistical Physics. Therefore we are not in presence of a real critical behavior characterized by universal power law exponents. The fact that in some cases one can fit the experimental data with power laws can be easily understood by considering that different agents might operate at different time scales. This would lead to a superposition of finite size effects corresponding to different time scales which may appear as a sort of effective power law exponent. The possibility that a suitable coupling between different time scales exists leading to genuine critical behavior is of course open. However, this is not the case for the class of model we are considering. In this perspective the variability of the effective exponents and their breakdown observed in various data [18] can be a genuine effect not simply due to limitation or problems with a database. This of course changes the perspective of the data analysis and of their comparison with the models. We expect that the behavior of different markets is reasonably similar because the key elements are essentially the same but without a strict universality.

In the asymmetric case one must consider two different temporal scales, say T_1 and T_2, the first referring to the formation of the chartists' bubble and the second relative to its duration. We have investigated with numerical simulations their dependence on N and δ, finding, as expected, a divergence of T_1 in the limit $N \to \infty$. Therefore the introduction of an asymmetry in the agents dynamics does not change the finite size effect associated to the quasi-critical behavior and the Stylized Facts.

4.3 Stylized Facts from the Model

In this section we discuss the results of some simulations of the minimal model described in Sect. 4.1. In order to clarify all the elements of the model we are going to discuss increasingly complex cases.

4.3.1 Single Agent

Considering that in the simplest model all agents are statistically identical (no real heterogeneity) we can fix our parameters in such a way that even a single stochastic agent can lead to an interesting dynamics. This single agent can be chartist or fundamentalist and the herding term is not active in this case. For simplicity we have also neglected the exponential term related to the price behavior in the transition probabilities. When the agent switches her behavior from chartist to fundamentalist, the market dynamics is in turn given by Eq. 4.1 or 4.3. If the agent is chartists she follows the market trend and creates bubbles, in this case the price fluctuations are larger with respect to a simple random walk. Otherwise if she is fundamentalist the price is driven towards the fundamental price and the fluctuations are smaller than the random walk ones. In this case the price is not diffusive and it remains almost constant, oscillating around the fundamental price p_f. In Fig. 4.5 we show the results of the simulation for the one-agent model. We can observe that the price dynamics displays local bubbles corresponding to periods in which the agent is a chartist. The price is instead almost constant when the agent is fundamentalist. Also we can observe periods of high or low volatility depending on the strategy of the agent. This dynamics clearly leads to fat tails in the distribution of price increments and also to a certain volatility clustering that certainly in this case is not due to the herding dynamics. This simple example shows that, once we have a full control of the model parameters, we can trace and reproduce some Stylized Facts even with an extremely minimal model.

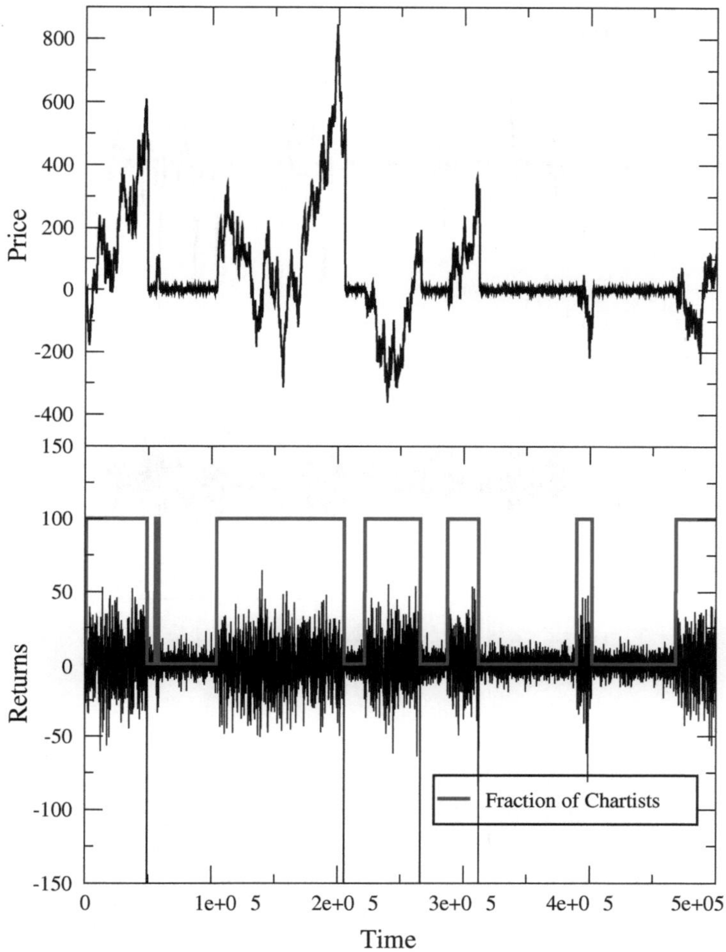

Fig. 4.5 Even a one agent model subjected to the stochastic dynamics can lead to non trivial results in terms of fat tails and volatility clustering. Here we can see the corresponding price and returns fluctuations. In this case we have also simplified the probability to change strategy by neglecting the exponential term corresponding to the price behavior. In the figure below the large returns are clearly related to the periods in which the agent is a chartist. The price behavior may look unrealistic because we consider a case with a fixed value of p_f. If one would also introduce a random walk behavior for p_f the price behavior would look much more realistic

4.3.2 Many Statistically Equivalent Agents ($N = 100$)

We now consider the more realistic dynamics with a larger number of agents. In principle we can tune the parameters to obtain the Stylized Facts for any preassigned value of N. For example in Fig. 4.6 (upper) we show the dynamics of the case

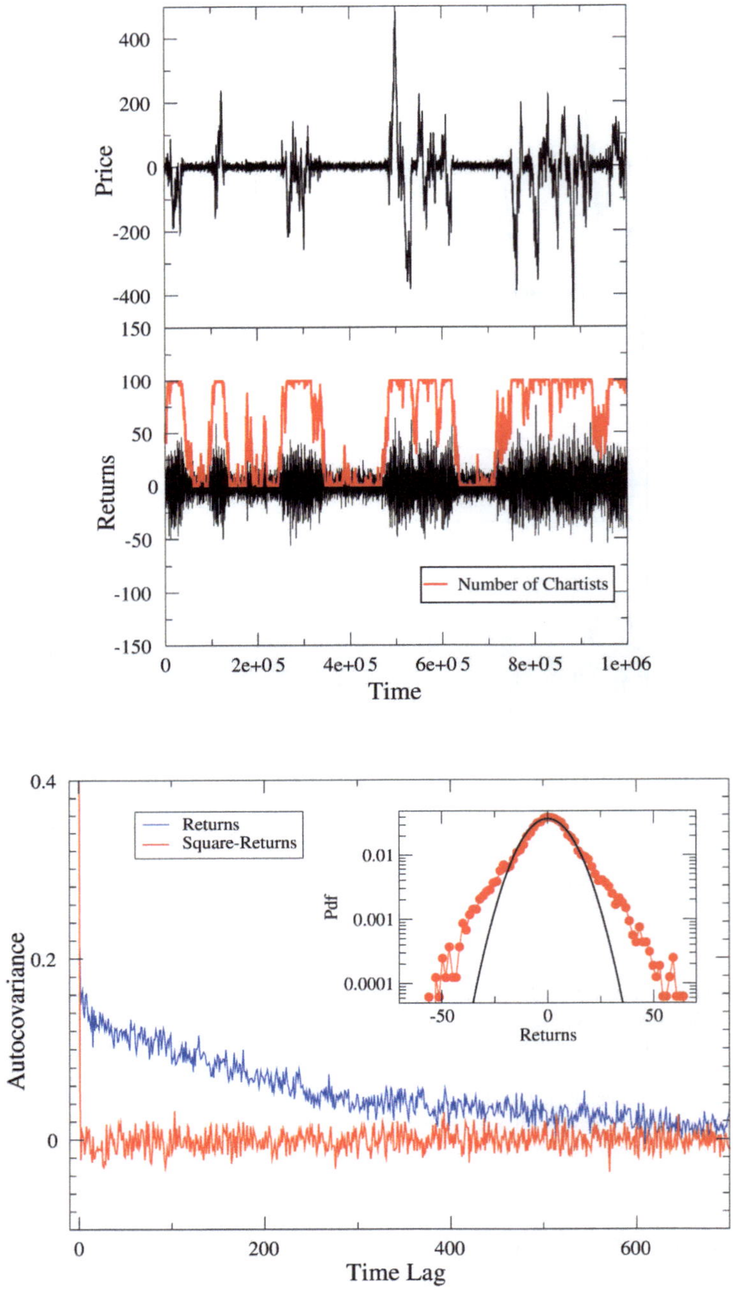

Fig. 4.6 We can adjust the parameters of the model in such a way that the Stylized Facts appear for any preassigned number of agents. The present case refers to $N = 100$. Also in this case we have omitted the exponential term of the price behavior in the probability to change strategy

$N = 100$ still without the exponential price term. Also in this case periods of high or low volatility correspond to regions in which chartist or fundamentalist agents dominate. We have analyzed the Stylized Facts for this model and both fat tails and volatility clustering are shown in Fig. 4.6 (lower). The square price-fluctuations shows a positive autocorrelation which (unlike real data) decays exponentially. As we have discussed this depends on the fact that this model has a single characteristic time scale.

4.3.3 Many Heterogeneous Agents

The limitation of a single time scale can be easily removed by introducing a real heterogeneity in the time scales of the agents' strategies. In particular we have introduced a distribution of values for the parameter M which is the number of steps agents consider for the estimation of the moving average of the price. We adopt a distribution of values also for the parameter b which gives the strength of the action of chartists. In this case we also introduce the exponential price term in the rate equations in order to have the most complete version of our model. In Fig. 4.7 we report the Stylized Facts corresponding to a uniform M distribution between 10 and 50 time steps and a uniform b distribution between 0 and 2.

The comparison of these results with those of the simplified model discussed before permits to draw the following conclusions:

- The exponential term related to the price behavior in the transition probabilities enhances the bubbles and crashes corresponding to the chartists action and it amplifies the fat tail phenomenon as shown by comparing Fig. 4.7 (lower) to Fig. 4.6 (lower). This is a qualitative argument because the two models are actually different and a precise comparison keeping the same effective parameters is not possible.
- The real heterogeneity of the agent time scales produces in this case a distribution of transition rates which leads to a quasi-power law behavior for the volatility correlations as shown in the inset of Fig. 4.7 (upper). On the other hand this different time scales have the effect of decorrelating the agents dynamics and the overall amplitude of the volatility is reduced.
- The short time anti-correlation for the returns (Fig. 4.7 upper), can be related to the long time predominance of fundamentalist (more details can be found in [19]).

4.4 Microscopic Origin of the Stylized Facts

As pointed out before, the intermittent behavior that characterizes the oscillations of agents' strategies is crucial to generate the Stylized Facts. One must choose a certain temporal window a correct value of N to obtain the "right" amount of fluctuations

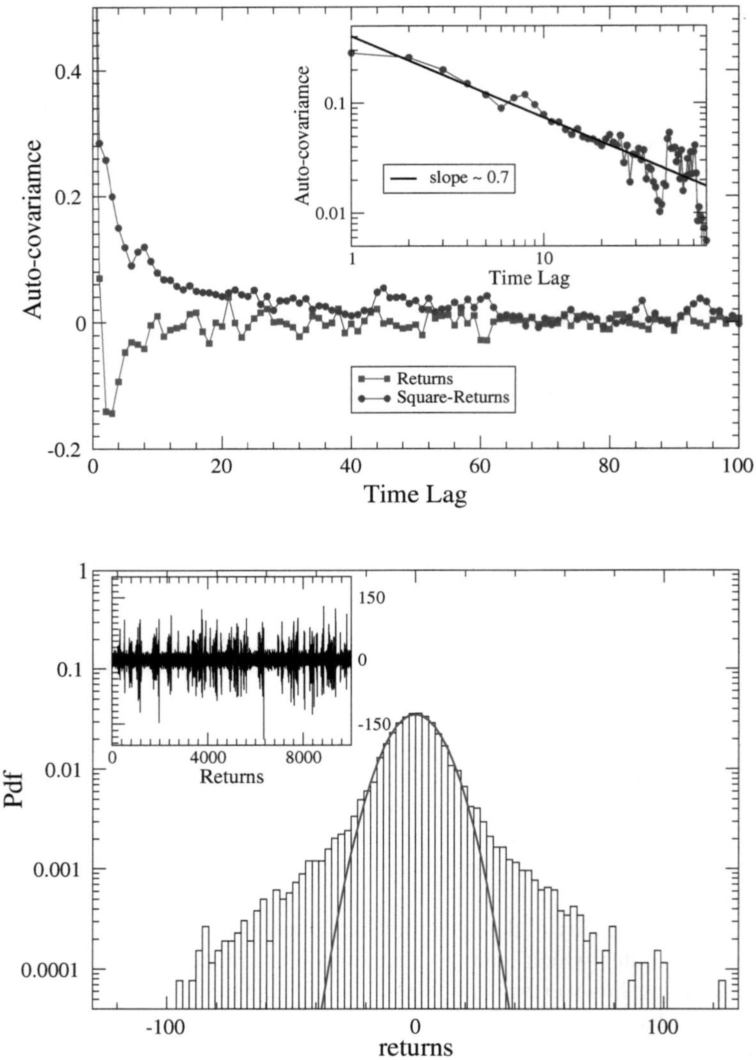

Fig. 4.7 Here we show the results for the most complete case we consider and in this case we adopt $N = 500$. This corresponds to including the exponential price term and introducing a real heterogeneity in the agents respect to their time horizons and impact on the market (see text discussion for details). We can see that the heterogeneity suppresses somewhat the volatility clustering because of a lack of coherence between the agents behavior but the shape of the volatility time correlation (see insert) in this case resembles power law. The exponential price term increases the size of the fat tails as shown in the lower figure

(the other parameters being fixed). In fact, while a low value of N produces too many fluctuations, a high value of N will prevent the formation of the chartists bubble at all. Therefore the Stylized Facts in our model can be regarded as finite size effects, that is, they disappear in the thermodynamic limit. Given the general nature of these results we may conjecture that it should apply to a broad class of agent based model and it is not a special property of our specific model.

This has important consequences for both data analysis and for the structure of the models proposed to investigate the origin of Stylized Facts. The fact that the exponents of the power laws characterizing financial markets don't seem to be universal [18] can be easily explained in this framework. The apparent power laws observed in many data find a natural explanation in the presence of a distribution of agents' strategies in terms of their time horizons and strength. Concerning universality, even if this is not strictly present in this model, it is reasonable to expect a certain similarity in all markets. This is due to the fact that the key elements to generate the Stylized Facts are of very general nature as we are going to explain in the following discussion.

The simplicity of the model permits to interpret the origin of Stylized Facts directly from the agents' strategies in the market. To this purpose it will be useful to define the concept of *effective action*. In our model agents not only decide between selling and buying, they operate in the market proportionally to the *signal* $p - p_f$ or $p - p_M$, selling or buying quantities which are proportional to this signal. Besides this, while in Statistical Mechanics the study of the dynamics of a model is usually done with a fixed value of N (and eventually infinite), nothing in financial markets permits us to justify this assumption. An agent can either enter or exit from the market on the basis of various considerations, and in addition she can varies the volume of the exchanges in the market. In summary, we must consider the effective action present in the market as the sum of these two effects, the fact that an agent can operate with different volumes and the fact that agents can enter or exit from the market.

Let us now suppose that for some reason at time t there is a price fluctuation (in any direction) Δp. Following our line of reasoning, this will produce an increase of the effective action, because both fundamentalists and chartists will see a signal in the market, and this action will produce more price fluctuations, and so on. On the contrary, if the market doesn't show opportunities to be profitable, agents will be discouraged from operating and so periods of low fluctuations will be followed by other periods of low fluctuations. This mechanism resembles the GARCH process, in which the volatility at time $t + 1$ depends on the volatility at time t and also on the return at time t [20]:

$$\sigma_{t+1} = f(\sigma_t; \Delta p_t). \tag{4.27}$$

We recall that the GARCH process was originally proposed as a phenomenological scheme to reproduce the phenomena of volatility clustering and fat tails. In our case we propose a microscopic interpretation of something similar to the GARCH process which, however, is now related to the specific agents dynamics.

On the other hand our model doesn't show any appreciable linear correlation between price increments, even if the arbitrage condition is not explicitly implemented in the model. This can be understood in the following way. Consider a price increment (with sign) Δp at time t. The next price increment will depend not only from the previous fluctuation, as it is for volatility, but also on all the other variables of the system, like the number of chartists, the specific values of p_M and p_f and so on. Schematically we can write:

$$\Delta p_{t+1} = f(\sigma; N_c; N_f; p_M; p_f). \tag{4.28}$$

All these additional variables are in general not correlated in a direct way with the price fluctuations, so they lead to a decorrelation of the price increments.

This line of reasoning explains qualitatively the presence of volatility clustering and the absence of linear correlations in financial markets. We believe that these considerations are of general nature and therefore they should be valid for all markets and models. Within our class of models, however, this general behavior does not reach the status of *universal behavior* in the sense of Statistical Physics.

4.5 Self-Organized Intermittency

In the previous section we have seen that our model is able to generate the Stylized Facts of financial markets and it is possible to control their origin and nature in great detail. However, in our model, as in most of the models in the literature, these Stylized Facts occur only in a very specific and limited region of the model parameters. This is a problem which is seldom discussed in the literature but in our opinion poses a vary basic question: why the market dynamics evolves spontaneously, or self-organizes, in the specific region of parameters which corresponds to the Stylized Facts?

In usual Critical Phenomena of Statistical Physics there is a basic difference between the parameters of the model (usually called coupling constants) and the number N of elements considered. Models with different coupling constants, but belonging to the same universality class, evolve towards the same critical properties in the asymptotic limit of very large N and very large time. This situation is called universality and it is often present in equilibrium critical phenomena in which the critical region requires an external fine tuning of various parameters. This is a typical situation of competition between order and disorder which occurs at the critical point.

Self-Organization instead occurs in a vast class of models characterized by a non linear dissipative dynamics far from equilibrium. It is important to note that this Self-Organization is also an asymptotic phenomenon, in the sense that it occurs in the limit of large N and large time.

For financial markets the possible analogy with SOC phenomena is very tempting. However, there are basic fundamental differences with the above concepts which require the developments of a different theoretical framework.

A very important observation is the fact that the Stylized Facts appear only for a specific value of the number of agents N. This result may appear as problematic in view of the above discussion about Self-Organization. A finite value of N cannot lead to universality and so the presence of the Stylized Facts in virtually all markets (with very different number of agents) appears rather mysterious. In this perspective a situation in which the Stylized Facts are generated only in the limit of large N would have appeared more natural. Also our model shows clearly that the intermittent behavior corresponding to the Stylized Facts must necessary occur for a finite value of N. It is easy to realize that this conclusion has a general nature with respect to the four essential ingredients of our class of models. An additional problem, in trying to explain the Self-Organization of the Stylized Facts, comes from the fact, that in addition to N, the model contains several other parameters. So, even if one would be able to produce the Stylized Facts in the large N limit, the Self-Organization with respect to the other parameters would remain unclear.

From our studies we propose a conceptually different mechanism of the Self-Organization phenomenon. Consider a certain market with its characteristic parameters. This values will govern the rate equation of the agents dynamics as described by Eqs. 4.4 and 4.5 and in Fig. 4.4. For any set of these parameters there will be a characteristic value of N which separates the regions with large fluctuations with the ones with small fluctuations, as shown for example in Fig. 4.4. The Stylized Facts appears precisely around this finite value of N, which corresponds to the right amount of intermittency. So our basic question about the Self-Organization is now transformed in a question related to the number of agents. Why the system should have precisely this number of agents acting on the market?

This question can now be answered in the following way. We have seen that if N is very large the system gets locked in the fundamentalist state and this leads to a very stable dynamics of the price. Such a situation will produce very small signals for the agent strategies. If we assume that an agent operates in an certain market only if her signal is larger than a minimum threshold, it is easy to realize that in the stable price situation, corresponding to large N, the number of active agents will decrease. On the other hand in the case of very few agents, the price dynamics undergoes to very large movements [21] and this produces large signals for the agents and will attract more agents in the market. Mathematically this concept can be implemented by introducing a threshold on the large scale price movements which corresponds to the decision of agents to be active or not active in the particular market. We have introduced this threshold in our model and a typical result is shown in Fig. 4.8. We can see that the variable number of agents evolves spontaneously towards the characteristic value of N corresponding to the intermittent state and the Stylized Facts. Since this state is not precisely critical in the sense of Statistical Physics we propose to call this phenomenon Self-Organized Intermittency. The thresholds are computed by analyzing the large scale fluctuations of the market price which we identify like price movements. This threshold mechanism does not imply anything about the small scale behavior of agents. In fact, it is possible to see that, by introducing a small scale threshold which discourages agents to enter in a noisy highly fluctuating market, the same results for the Self-Organization can be recovered.

This new vision has important implications in various directions. On one hand it seems to be a reasonably natural and robust explanation of the Self-Organization towards the quasi-critical state, on the other hand the quasi-critical state leading to

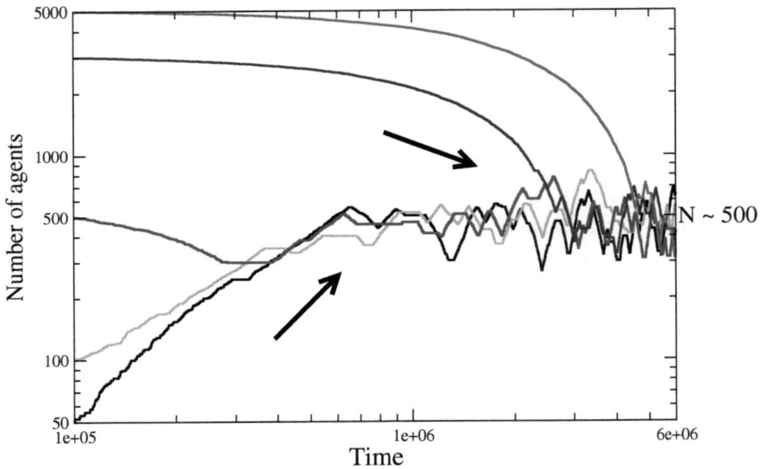

Fig. 4.8 Self-Organized Intermittency. We can observed the self-organized nature of the dynamics toward to the quasi-critical state corresponding to existence of stylized facts. We show five cases, characterized by different initial conditions, $N(t = 0) = 5000, 3000, 500, 100$ and 50. If the number of the agents in the market is too large ($N = 5000, 3000$) the price will be stable and essentially driven by fundamentalists. In this case the market is not profitable and agents will not operate. So N will decrease. On the other hand a too small number of agents ($N = 50, 100$) will lead to large fluctuations and will attract more agents, increasing the value of N. In all cases the system converges to an intermediate situation ($N \approx 500$ in this case) which corresponds to the intermittent behavior which leads to stylized facts. Note that this regime is not real critical and universal in the sense of statistical physics

the Stylized Facts is unavoidably linked to finite size effect. A first consequence is that in the limit of large N, or in the limit of large time, the Stylized Facts disappear. A fact which seems to be indeed reproduced by price time series [18]. Another consequence is that genuine critical exponents and universality can not be expected in this framework. The fact that several datasets can be fitted by power laws is however not surprising. Since different agents may have different time horizons in their trading strategies, one has a superposition of finite size effects which may appear as a power law in some range. This perspective implies that the discrepancy in the effective exponents for different dataset and for different time scales [18] may be a genuine intrinsic property instead of spurious effect due to the incompleteness of the data.

A key element of the present scheme is the nonstationarity of the system with respect to the value of N which represent a major departure from usual critical phenomena. On the other hand this non stationarity appears to be a very important element in real financial markets.

4.6 Generalization of the Models

The main results of our model are:

- Detailed understanding of the origin of the Stylized Facts with respect to the microscopic dynamics of the agents.
- Demonstration that in this class of model the Stylized Facts correspond to finite size effects and not to universal critical exponents. This finite size effect, however, can be active at different time scales.
- Bubbles of chartists can be triggered spontaneously by a multiplicative cascade which can originate from tiny random fluctuations. This situation resembles in part the avalanches of the Sandpile Model in Statistical Physics [22].
- We have shown the importance of the non stationarity in the dynamics of the number of active agents N and we introduced a characteristic threshold to decide when an agent can enter or exit from the market.
- This threshold and the relative non stationarity are proposed to represent the key element in the self-Organization mechanism. This Self-Organization, however, leads to an intermittency related to finite size effects. For this reasons we define it as Self-Organized Intermittency.

Starting from the minimal model introduced in this chapter and considering that one can obtain a detailed microscopic understanding of its dynamics, it is easy to identify a number of realistic variants which can be introduced as generalizations of the model (this indeed holds also for the investment rules introduced to explain the Self-Organization of financial Markets).

4.6.1 Multiplicative Dynamics

In the chapter [19] (we only give some hints of the analysis carried out in this work) we focus on various statistical properties from both a numerical and analytical point of view. We will also discuss in some detail the differences between linear and multiplicative dynamics for the price behavior while In the previous sections we have only considered the linear dynamics for simplicity.

The role of Walras' price adjustment mechanism (see [13]) in economics is very similar to the role of Newton's law of the dynamics in physics because Walras' mechanism introduces a relationship between the excess demand (ED) and the consequent variation of the price that ED causes (the excess demand is usually defined as a function of the unbalance of demand and supply). This law has been formalized in many ways since it was introduced about two centuries ago and the version, which belongs to the framework of our model, is the following

$$\frac{1}{p}\frac{dp}{dt} = \beta\, ED \tag{4.29}$$

or when time is discrete ($\Delta t = 1$):

$$\frac{\Delta p}{p} = \beta\, ED \tag{4.30}$$

The basic idea of Eq. 4.29 is that price follows a multiplicative process rather than a linear one so its percentage increments, rather than the increments themselves, are proportional to ED. There is appreciable evidences in the experimental data that a multiplicative dynamics is indeed more appropriate than a linear one [23, 24].

Up to now we have neglected this multiplicative nature in our model and we have adopted the simplified linear dynamics. In fact the main purpose of our model is to trace back the detailed origin of the Stylized Facts and their self-organization. In this perspective the linear dynamics does not change the essence of these elements but it represents an useful simplification especially in deriving some analytical results. Here we briefly discuss in detail the analogies and differences of the two regimes. It is important to underline that the (simplified) population dynamics adopted in the multiplicative case is the same of the linear regime.The motivation is due to the fact that the exponential term, associated to the price and present in the complete version of the transition probability, would be always dominant with respect to the herding term because of the great variability of the price in the multiplicative version.

In our model the (linear) price increments are proportional to the term $-x\, b/(M-1)(p_t - p_M) + (1-x)\gamma(p_f - p_t)$ which we identify as the ED, in this respect it corresponds to a linearized Walrasian mechanism. Considering this as the small price increments limit of a multiplicative dynamics we can now try to go back and consider the real dynamics from which this approximation may arise. In order to construct a price equation which is consistent with Eq. 4.30, we define the price increments as a geometric random walk with an additional term that is indeed the excess demand:

$$\ln(p_{t+1}) = \ln(p_t) + ED + \sigma \xi_t \tag{4.31}$$

where ξ_i are independent and normally distributed random variables. However the walrasian mechanism does not specify which is the correct choice for ED, apart from the fact that the dimensional analysis of Eq. 4.30 implies that ED must be somehow normalized (dimensionless). In our linear version (Eqs. 4.9 and 4.10) the term that plays the role of ED has the dimension of a price so the simplest way to make the ED of the linear version dimensionless is to normalize it with a quantity which has the dimension of a price. This normalization is not univocal and we choose (see [19] for more details about this choice)

$$ED = x\frac{b}{M-1}\frac{(p_t - p_M)}{p_t} + (1-x)\gamma\frac{(p_f - p_t)}{p_t} \tag{4.32}$$

with obviously $p_f \neq 0$. Most of the ABMs share the common aspect that they (more or less) reproduce the empirical facts (i.e. Stylized Facts) only in a certain region of the phase space of their parameters. A typical example is the model of Lux and

Marchesi [2, 3]. Therefore they require a fine tuning of all the parameters and even of the total number of agents N in order to reproduce the Stylized Facts.

This situation is already present in the models with the linear dynamics. The multiplicative version, as for example in our model Eq. 4.31, is dramatically unstable and more sensitive to small changes of parameters than the linear one. The results is that the dynamics of price cannot be simulated for a large part of the phase space of the parameters because it is subject to extreme singularities towards zero or infinity. Consequently the main attempt of tuning the parameters consists in finding the subtle equilibrium between the weight of the noise (its variance σ) and the weight of the ED (b and γ) in the evolution Eq. 4.31 of the price.

We have argued that the linear version should be a linearization of the original Walrasian mechanism for price adjustment Eq. 4.30. If we now linearize Eq. 4.31 with the normalization of Eq. 4.32, it follows that (Fig. 4.9)

$$p_{t+1} - p_t \approx (\sigma p_t)\xi_t + x\frac{b}{M-1}(p_t - p_M) + (1-x)\gamma(p_f - p_t). \qquad (4.33)$$

This linearization is justified if $(p_{t+1} - p_t)/p_t \ll 1$ and this condition is generally true for our choice of parameters. The comparison of Eq. 4.33 with Eqs. 4.9 and 4.10 highlights that the only difference lies in the variance of the white noise. This variance is a random variable in Eq. 4.33 and the linear version Eqs. 4.9 and 4.10 is essentially the linearization of the multiplicative dynamics if the variance of the white noise can be assumed as nearly constant neglecting its stochastic nature. Let us define the effective variance as $\hat{\sigma} = \sigma p_t$, the random variables ξ_t and p_t are

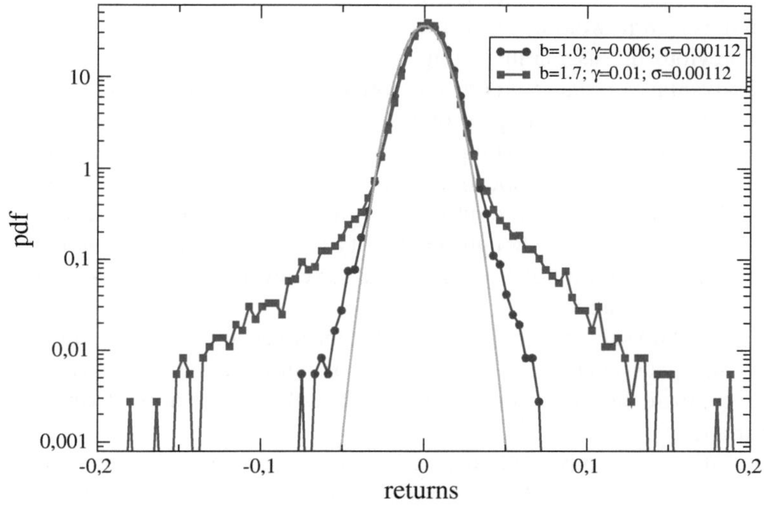

Fig. 4.9 The multiplicative process results appear to be extremely sensitive of the region of parameters chosen. Here we show two slightly different set of parameters which lead to drastically different results with respect to the fat tails phenomenology

Table 4.1 Estimation of Ω for different value of p_f

p_f	Ω
1	0.029
10	0.031
50	0.032
100	0.034
500	0.034
1000	0.035

independent because $p_t = p_t(\xi_0, \xi_1, \ldots, \xi_{t-1})$, hence we can focus just on the effective variance $\hat{\sigma}$. In the average $E[\hat{\sigma}] = p_f \sigma$ and the typical fluctuation Ω, normalized with the squared average, is

$$\Omega = \sqrt{\frac{E[p_t^2]}{p_f^2} - 1} \approx \sqrt{\frac{E[p^2]}{p_f^2} - 1} \tag{4.34}$$

The linear process of Eqs. 4.9 and 4.10 is therefore justified if the fluctuations given by Eq. 4.34 are negligible. In other words Eqs. 4.9 and 4.10 are the limit of the linearized Eq. 4.33 when Ω goes to zero. In Table 4.1 we show that fluctuations around the mean value of $\hat{\sigma}$ are very small independently on p_f. This shows that for a choice of the parameters which appears reasonable to reproduce real market features, this effect is only of the order of 3 %. Therefore the linearized version of the dynamics indeed is a valid approximation to the multiplicative one. It is important to note however that a different choice of the parameters could lead to a situation in which even in the limit of small price increments the simple linearized (with the random noise) form would be incorrect.

The previous discussion may suggest that the multiplicative version introduces only disadvantages because it is much more unstable in comparison with the linear version with respect of phase space of the parameters. In general very few results can be derived analytically for the complete model. But it have also important advantages, the first one is that the multiplicative dynamics is more realistic. The Stylized Facts of the multiplicative version are substantially more pronounced with respect to those of the linear version. Even though stationarity is certainly weaker than in the linear dynamics also the multiplicative case appears to converge to a quasi-stationary state but with larger fluctuations.

In Fig. 4.10 we show the return pdf for the linear version and for the multiplicative one. The non gaussian tails are much more pronounced in the multiplicative case. In order to compare the two cases we have rescaled the returns normalizing them with their variance.

In Fig. 4.11 we compare the normalized autocovariance of returns and squared returns for the two cases. The two models give similar results from a qualitative point of view but the degree of correlation of squared returns is nearly twice in the multiplicative case. Volatility clustering is therefore enhanced by the multiplicative dynamics.

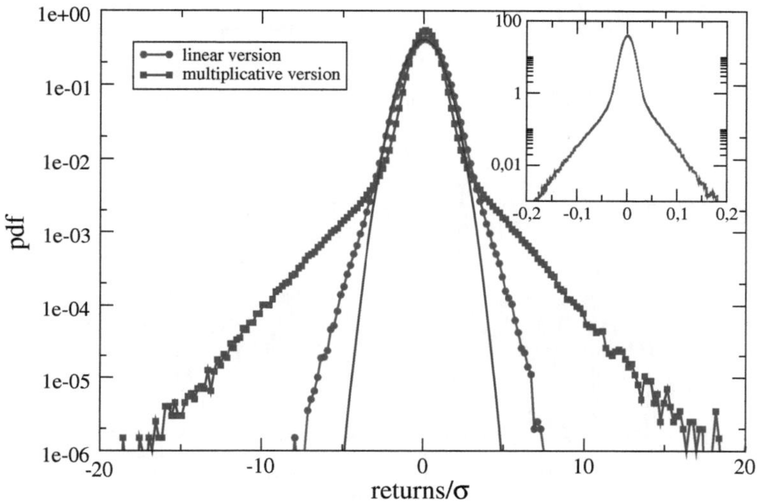

Fig. 4.10 Fat tails for linear and multiplicative dynamics (time interval equals 100 steps). The multiplicative case is characterized by much stronger non gaussian tails (the returns are normalized with their variance). The solid line represents a gaussian plotted for comparison. The small panel on the right shows the non normalized returns probability density function in the multiplicative case

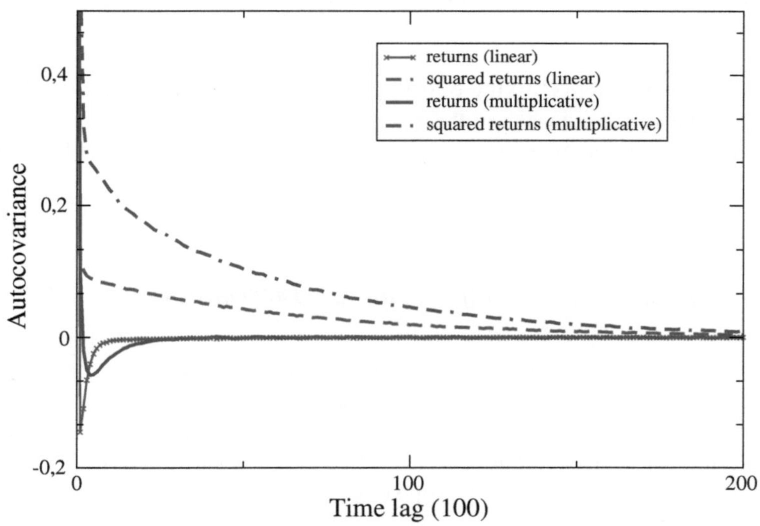

Fig. 4.11 Volatility clustering and market efficiency for linear and multiplicative dynamics. The autocovariance function of returns and of squared returns are qualitatively similar in the two cases. The volatility clustering is larger in the multiplicative case than in the linear one. The linear correlation shows very similar behavior. Note that in this minimal version of the ABM both the linear and multiplicative dynamics lead to a single characteristic time scale and therefore volatility clustering has an exponential decay. The possibilities of multiple time scales for the agents can modify this behavior towards a more realistic one (power law like) as shown in Fig. 7 of [25]

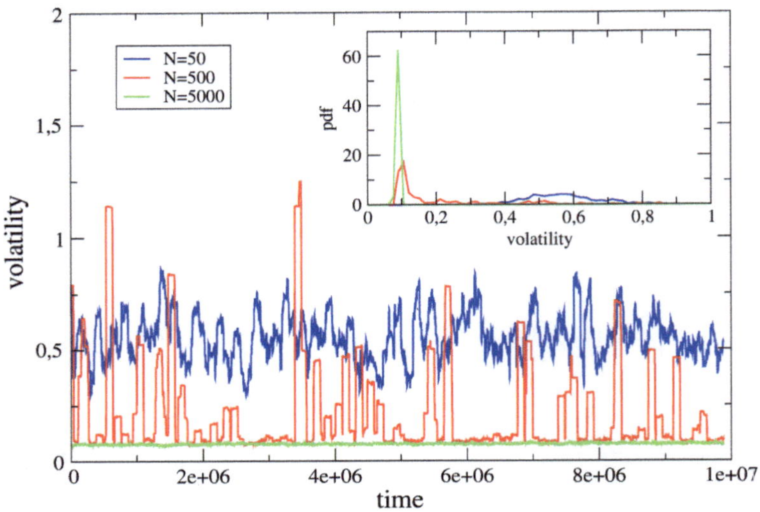

Fig. 4.12 Volatility fluctuations for various values of N. As in [21, 25] we see that for large values of N ($N = 5000$), the volatility is very small and this situation is not interesting for the agents. For small values of N ($N = 50$), instead, the volatility is always very high and the market offers arbitrage opportunities for the agents. If $N = 500$ we observe an intermittent behavior as in real market. The inset reports the histogram of the main plot

In the end, with respect to the Self-Organization of markets, the overall picture is very similar to the linear case with the only difference that the convergency to the quasi-critical is somewhat slower with respect to the linear case for large N (see Figs. 4.12 and 4.13).

4.6.2 Robustness of the Self-Organized Intermittency

In Sect. 4.5 we have seen that, by fixing the thresholds Θ_{in} and Θ_{out} in a region of values of the volatility σ which is intermediate with respect to the two limit cases $\sigma_{N=5000}$ and $\sigma_{N=50}$, the system self-organizes in the intermittent state corresponding to an intermediate number of agents $N^* \simeq 500$ (see Figs. 4.12 and 4.13).

In doing this we have chosen the threshold values to correspond approximatively to the region of the fluctuations leading to the Stylized Facts. This may induce the idea that by choosing different values of Θ_{in} and Θ_{out} one may force the system to self-organize to any preselected state, not necessarily the one corresponding to the Stylized Facts. This is not the case because by choosing unreasonable thresholds' region the system does not reach an interesting or unique self-organized state. In Fig. 4.14 we have plotted the histograms of the volatility σ corresponding to three different populations with a fixed number of agents N with values 50, 500, 5000.

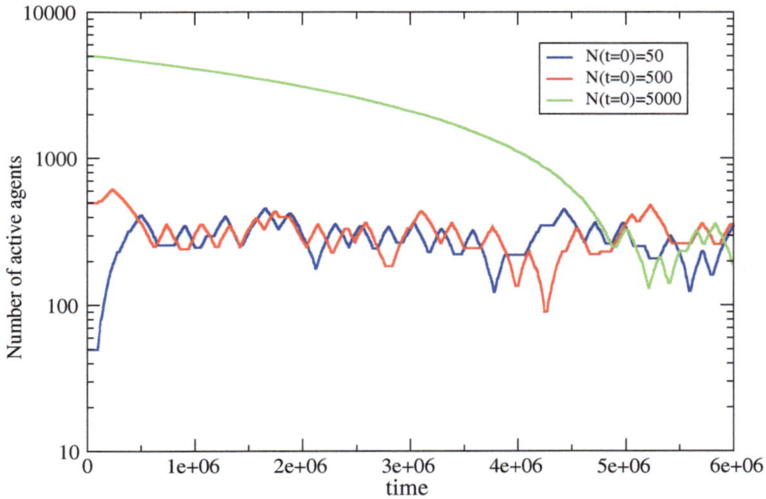

Fig. 4.13 Self-Organized Intermittency. The mechanism of self-organization is the same of the linear case proposed in [21, 25]. When N is larger than N^* ($N^* \approx 500$ is the number of agents that shows SF) fluctuations are typically lower than Θ_{out} therefore agents leave the market. Instead when N is smaller than N^* the fluctuations are larger than Θ_{in} and the agents enter into the market

We can see that the volatility of the $N = 5000$ population is very sharp and it is picked around a small value of σ. In the case of $N = 50$ the histogram is broader and has a maximum on a very high level of volatility. The situation is different in the intermediate case of $N = 500$ where the distribution in very broad and asymmetric. It is picked on small values of volatility but the tails reach very high values, much more than the $N = 50$ population. The reason for the high values of price fluctuations of the intermediate case ($N = 500$) with respect to the extreme case ($N = 50$) is the following. A very large price fluctuation corresponds to a situation in which the chartists action can develop for a certain time. In the highly fluctuating regime ($N = 50$) the life time of chartist action is too small for this to happen. On the contrary for the intermediate case chartist fluctuations are more rare but when they happen they may last for a longer time.

We now consider three different possibilities for the thresholds values of Θ_{in} and Θ_{out}. These regions are evidenced in Fig. 4.14 with different colors. The first region is centered on low values of volatility, the second on high values and the third on intermediate values. By choosing this last region, that is the one used in Sect. 4.5, the system self-organizes in the intermittent state which corresponds to a fluctuating intermediate number of agents ($N \simeq 500$). Instead unrealistic anomalies occur if one chooses the other two regions. When the region defined by Θ_{in} and Θ_{out} is centered on low volatility values, and the starting number of agents is small, the system size grows until it reaches a number of agents which leads to an average value of the volatility σ which is inside the region considered. Then the fluctuations

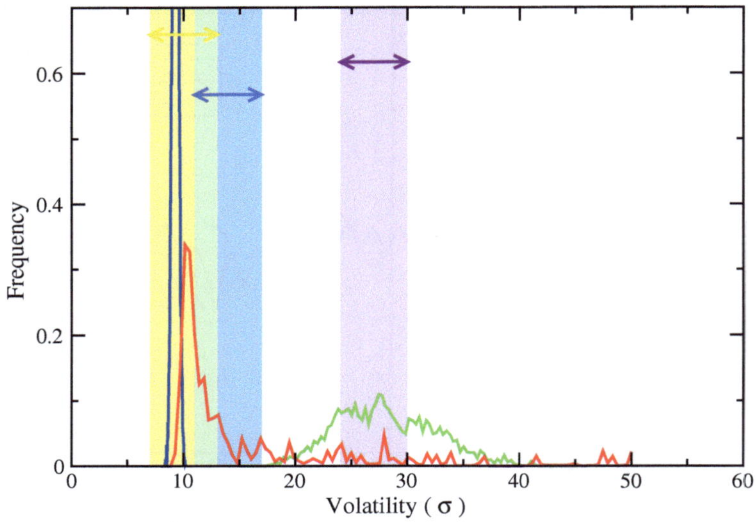

Fig. 4.14 Histograms of the volatility σ for different populations with fixed number of agents N. The picture refers to three populations with N equal to 50 (*green line*), 500 (*red line*) and 5000 (*blue line*). The histograms referring to $N = 50$ and $N = 5000$ are almost symmetric with the difference that the first is broad and the second very sharp. The histogram for $N = 500$ is asymmetric and has tails which extends for very large values of the volatility σ, even larger than the ones of the $N = 50$ histogram. By considering this plot, we have identified three different regions (indicated in the picture with different colors), delimited by different values of Θ_{in} and Θ_{out}, to trigger the self-organization towards different values of N. We will see that only by choosing a suitable region centered on the maximum of the histogram which refers to $N = 500$ one can obtain a market dynamics which self-organizes towards a stable value

from this average value are so small that the system is actually locked in a certain (high) value of the number of agents. The dynamics corresponding to this situation does not display the SF anymore. Of course when the starting number of agents is very large its average volatility level is always inside the considered region and the system size is constant in time. The dynamics corresponding to a low volatility centered region is shown in Fig. 4.15 (left) and it does not lead to the phenomenon of the self-organization.

On the contrary, as shown in Fig. 4.15 (right), when the region defined by the thresholds is centered on very high values of volatility the system size rapidly drops down because the system has an average volatility which is always smaller than the threshold considered to enter the market. In this way the system collapses to the unrealistic situation of zero-agent population where the price fluctuations are only due to the random noise term.

Therefore this study shows that the phenomenon of self-organization and the presence of stylized facts are intrinsically linked and one cannot force the system to self-organize to a reasonable dynamics which does not lead to the SF. For example this forbids the possibility of self-organization associated to a random walk dynamics (associate to the SF).

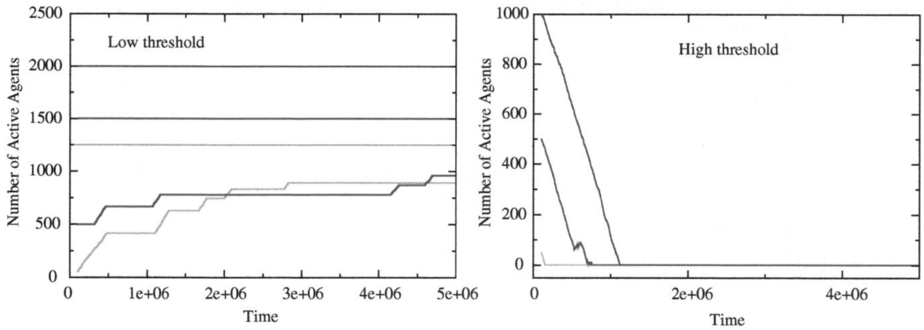

Fig. 4.15 Self-organization using other rules. In this plot we have analyzed the self-organization phenomenon using values of Θ_{in} and Θ_{out} which defines regions which are centered on low (*left plot*) and high (*right plot*) values of volatility. We can see that by choosing these regions the self-organization phenomenon is no more observed. In the left panel we can see that the number of agents $N(t)$ is an always increasing function which becomes a constant when $N(t)$ exceeds a certain value. The opposite happens in the right panel where $N(t)$ decreases step by step going towards the unrealistic situations with $N = 0$

The threshold mechanism could be apparently problematic because it may be argued that investors could be scared by a too fluctuating market [26]. However, this problem can easily clarified by the analysis of fluctuations at different time scales. The price movement which we interpret as a positive signal for the agents' strategy corresponds to the volatility at relatively long time scale. On the other hand a large volatility at a shorter time scale would induce a high risk on such a strategy. In this section we consider how this problem may affect the self-organization mechanism. The introduction of this more complex and realistic scenario in the model does not change the essential elements of the self-organization phenomenon. Since we want that agents look at fluctuations on two time horizons, at each time step t they now have to evaluate fluctuations $\sigma(t, T)$ for two different values of T that we call T_1 and T_2 corresponding respectively to the small time scale and to the large time scale. We set $T_2 = 1000$ as in the previous section and we choose $T_1 = T_2/100$.

The fear of a too volatile market at a short time scale can be represented by the new threshold Θ_s. If $\sigma(t, T_1) > \Theta_s$ the agent will consider the situation as dangerous and she will exit the market with a certain probability. If the agent is inactive and the previous condition is fulfilled she will not enter in the market. Instead if the opposite condition is true (i.e $\sigma(t, T_1) < \Theta_s$) the agents compare the long time scale fluctuations $\sigma(t, T_2)$ with the thresholds Θ_{in}, Θ_{out} and enter/exit according to the same scheme of Sect. 4.5. In Fig. 4.16 we can see that, independently on the starting number of agents, the system tends to the quasi-critical state (i.e. N^*) with the Stylized Facts as in the previous section for a suitable choice of the thresholds Θ_{in}, Θ_{out}, Θ_s. The unique effect introduced by Θ_s is a slight asymmetry between the rise and the decrease of the number of agents N. The value of Θ_s we adopt (Fig. 4.17) is quantitative of the same order of Θ_{in} and Θ_{out}, only it corresponds to shorter time scales.

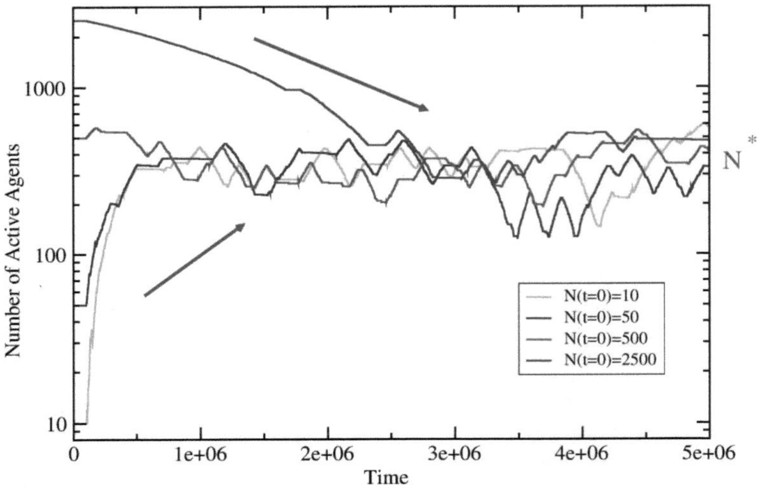

Fig. 4.16 Self-Organization with risk-scared agents. The introduction of a small-scale threshold Θ_s does not change the results in fact the system self-organizes into the quasi-critical state N^* independently on the starting number of agents. The unique effect introduced by Θ_s is a slight asymmetry between the rise and the decrease of the number of agents N

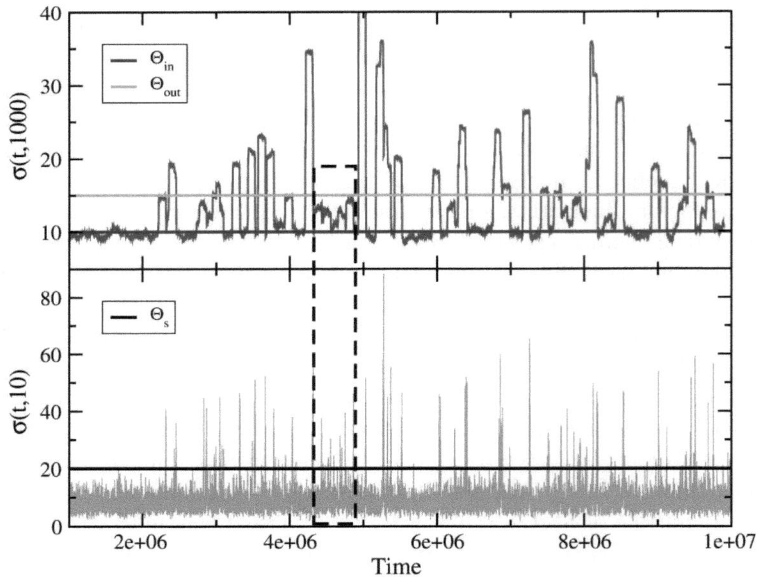

Fig. 4.17 Analysis of the volatility on different time scales. Large fluctuations of $\sigma(t, T_1)$ does not necessary imply large fluctuations of $\sigma(t, T_2)$ and vice-versa as it can be seen in the highlighted region. In fact while $\sigma(t, T_2) < \Theta_{in}$ we have at the same time that $\sigma(t, T_1) > \Theta_s$

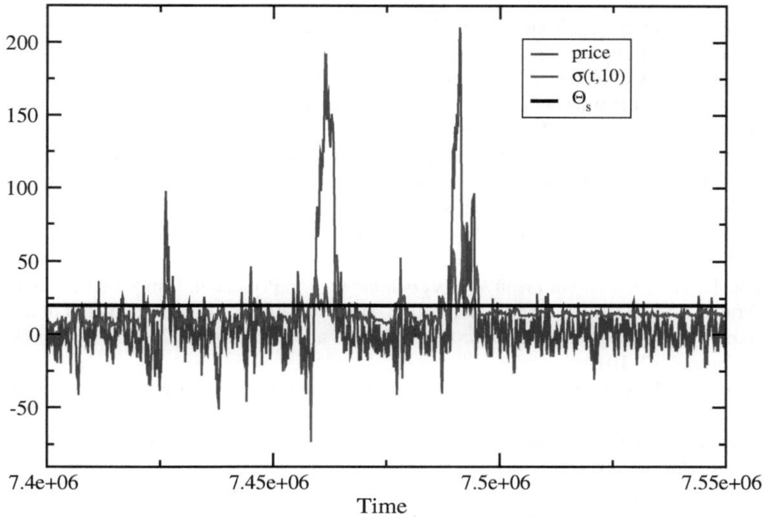

Fig. 4.18 Analysis of the short-scale volatility threshold. We see that the condition $\sigma(t, T_1) > \Theta_s$ is fulfilled when the price makes very large fluctuations and grows (or drops) very quickly on the scale of T_1

It is also interesting to note that the presence of large fluctuations on the scale of T_1 does not imply large fluctuations on the scale of T_2 or vice-versa as Fig. 4.17 points out. In fact we can see in the highlighted region that, while $\sigma(t, T_2)$ is smaller than Θ_{in}, $\sigma(t, T_1)$ is instead usually larger than Θ_s. To conclude this section we report in Fig. 4.18 the price behavior and the fluctuations $\sigma(t, T_1)$, the small scale mechanism is active only when the price has large fluctuations on the small scale.

References

1. Bachelier, L. (1900). Théorie de la spéculation. Annales Scientifiques de l'Ecole Normale Supérieure, *III-17*, 21. *Ph.D. thesis is mathematics.*
2. Lux, T., & Marchesi, M. (1999). Scaling and criticality in a stochastic multi-agent model of a financial market. *Nature*, *397*, 498.
3. Lux, T., & Marchesi, M. (2000). Volatility clustering in financial markets: A micro-simulation of interacting agents. *International Journal of Theoretical and Applied Finance*, *3*, 675.
4. Vaglica, G., Lillo, F., Moro, E., & Mantegna, N. (2008). Scaling laws of strategic behavior and size heterogeneity in agent dynamics. *Physical Review E*, *77*, 036110.
5. Takayasu, M., Mizuno, T., & Takayasu, H. (2006). Potential force observed in market dynamics. *Physica A*, *370*, 91.
6. Mizuno, T., Takayasu, H., & Takayasu, M. (2007). Analysis of price diffusion in financial markets using PUCK model. *Physica A*, *382*, 187. arXiv:physics/0608115.
7. Alfi, V., Coccetti, F., Marotta, M., Pietronero, L., & Takayasu, M. (2006). Hidden forces and fluctuations from moving averages: A test study. *Physica A*, *370*, 30.

8. Alfi, V., De Martino, A., Tedeschi, A., & Pietronero, L. (2007). Detecting the traders' strategies in minority-majority games and real stock-prices. *Physica A*, *382*, 1.
9. Takayasu, M., Mizuno, T., & Takayasu, H. (2005). Potentials of unbalanced complex kinetics observed in market time series. arXiv:arXiv:physics/0509020.
10. Watanabe, K., Takayasu, H., & Takayasu, M. (2009). Random walker in temporally deforming higher-order potential forces observed in a financial crisis. *Physical Review E*, *80*, 056110.
11. Alfi, V., De Martino, A., Pietronero, L., & Tedeschi, A. (2007). Detecting the traders' strategies in minority-majority games and real stock-prices. *Physica A*, *382*, 1.
12. Kirman, A. (1993). Ants, rationality and recruitment. *The Quarterly Journal of Economics*, *108*, 137.
13. Kreps, D. M. (1990). *A course in microeconomic theory*. Princeton: Princeton University Press.
14. Alfarano, S., Lux, T., & Wagner, F. (2008). Time-variation of higher moments in a financial market with heterogeneous agents: An analytical approach. *Journal of Economic Dynamics and Control*, *32*, 101.
15. Alfarano, S., Lux, T., & Wagner, F. (2005). Estimation of agent-based models: the case of an asymmetric herding model. *Computational Economics*, *26*, 19.
16. Gardiner, C. (1990). *Handbook of stochastic methods: for physics, chemistry and the natural sciences*. Berlin: Springer.
17. Egenter, E., Lux, T., & Stauffer, D. (1999). Finite-size effects in monte carlo simulations of two stock market models. *Physica A*, *268*, 250.
18. Cont, R. (2001). Empirical properties of asset returns: Stylized facts and statistical issues. *Quantitative Finance*, *1*, 223.
19. Alfi, V., Cristelli, M., Pietronero, L., & Zaccaria, A. (2009). Minimal agent based model for financial markets II: statistical properties of the linear and multiplicative dynamics. *European Physical Journal B*, *67*, 399. arXiv:0808.3565.
20. Bollerslev, T. J. (1986). Generalized autoregressive conditional heteroskedasticity. *Econometrics*, *31*, 307.
21. Alfi, V., Pietronero, L., & Zaccaria, A. (2009). Self-organization for the stylized facts and finite-size effects in a financial-market model. *EPL*, *86*, 58003.
22. Bak, P. (1996). *How nature works: The science of self-organised criticality*. New York: Copernicus Press.
23. Mantegna, R. N., & Stanley, H. (2000). *An introduction to econophysics: Correlation and complexity in finance*. New York: Cambridge University Press.
24. Bouchaud, J. -P., & Potters, M. (2003). *Theory of financial risk and derivative pricing: From statistical physics to risk management*. Cambridge: Cambridge University Press.
25. Alfi, V., Cristelli, M., Pietronero, L., & Zaccaria, A. (2009) Minimal agent based model for financial markets I: origin and self-organization of stylized facts. *European Physical Journal B*, *67* (2009), 385. arXiv:0808.3562.
26. Pagano, M. (1989). Endogenous market thinness and stock price volatility. *The Review of Economic Studies*, *56*, 2.

Part II
Financial Market Dynamics: Order Book and Data Analysis

Chapter 5
Order Book: Introduction and Main Statistical Evidences

The order book is the elementary mechanism of price evolution. In fact the order book is literally the book (now it is an electronic register) where the participants' orders are stored if they are not immediately executable. The order book is a double auction system which records the orders of selling and buying and, in case, matches compatible orders as we are going to see in the next section. In this chapter we describe the mechanism of working of the order books and then we review their main statistical properties.

5.1 Order Books in a Nutshell

The elementary mechanism of price formation is a double auction where traders submit orders to buy or sell. The complete list of all orders is called the *order book*. In Fig. 5.1 we give a schematic representation of an order book. There are two classes of orders: market orders and limit orders.

- **Market orders** correspond to the intention of immediately purchase or sale at the best price (quote) available at that time.
- **Limit orders** instead are not immediately executed since they are offers to buy or sell at a certain quote which is not necessary the best one. If we consider a sell limit order this means that its quote is higher than (or equal to) the best bid $b(t)$ which is the order of buying with the highest price. On the other hand a buy limit order implies that its price is lower than (or equal to) the best ask $a(t)$ which is the order of selling with the lowest price.

The non-zero difference between $a(t)$ and $b(t)$ is defined as the spread $s(t) = a(t) - b(t)$. The prices of placement of orders (called 'quotes') are not continuous but quantized in unit of ticks whose size is an important parameter of an order book.

The price of a stock can be conventionally defined as the mid-price $p(t) = [a(t) + b(t)]/2$ and it can change only if a limit order falls inside the spread or if a market order matches all the orders placed at the best quote. It is clear that the specific

M. Cristelli, *Complexity in Financial Markets*,
Springer Theses, DOI: 10.1007/978-3-319-00723-6_5,

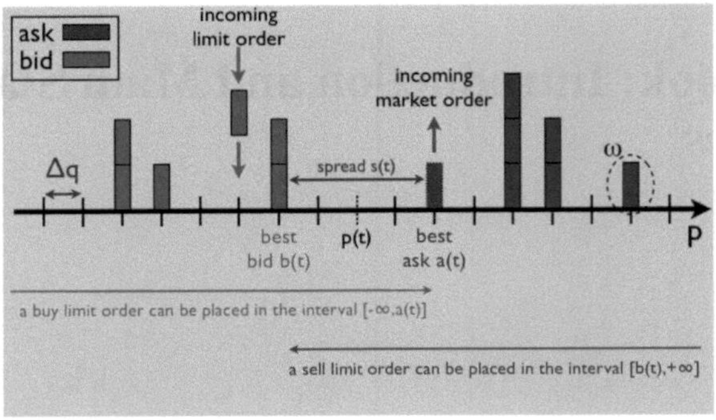

Fig. 5.1 Schematic representation of the order book dynamics. There are two classes of orders: the market orders and the limit orders. The market ones are orders of purchase or sale at the best available quote. On the other hand the limit orders are not immediately executed since they are placed at a quote which is less favorable than the best quote. The volume ω of the orders is an integer multiple of a certain amount of shares. The price of a stock can be defined as the mid-price $p(t) = [a(t) + b(t)]/2$

configuration of an order book it is a very important aspect for price movements. A thick book full of orders can absorb the arrival of new orders without giving rise to large jumps of the price. On the contrary if the book is sparse, even a small incoming order could trigger a large price variation. In order to clarify this problem we can consider an order book in the configuration of *panel a* in Fig. 5.2. This situation corresponds to a very liquid market in which the order book can absorb several orders without large price variations even if they are relatively large. This regime corresponds to the assumptions of the standard financial picture where an order can be immediately executed, its impact is marginal and the market is nearly efficient.

Instead if a liquidity crisis occurs, the configuration of the order book changes dramatically and the situation is like the one represented in *panel b* in Fig. 5.2. The orders stored are few, the average distance between them is large and the flow of market orders cannot be easily and immediately absorbed by the system and by consequence even a small order can produce a large price variation.

In the next sections we are going to give an overview to the main features and statistical evidences of the dynamics of real order books.

5.2 Order Books' Stylized Facts

The intensive and systematic investigation of order books started about fifteen years when a huge amount of empirical data produced by financial markets became easily available. Large data availability is a crucial point because Economics and in general Social Sciences set a further conceptual problem beyond the investigation of the Natural Laws itself: the existence or not of some underlying Natural Laws to investigate. In Physics the existence of these Laws is always assumed, instead in Economics this

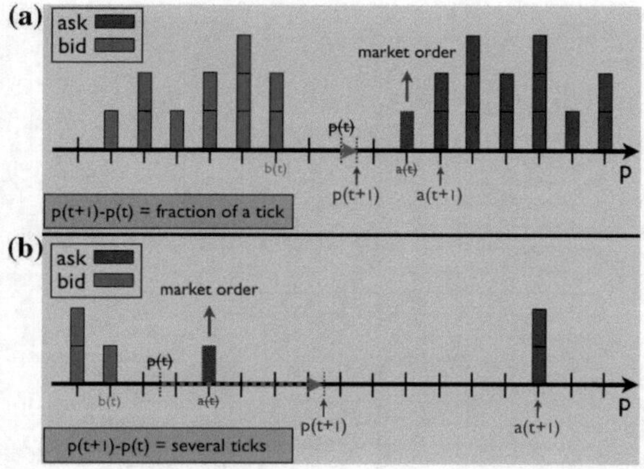

Fig. 5.2 A very liquid order book versus a liquidity crisis. This figure illustrates how the degree of liquidity of a market plays a key role in determining the system response to the volume of an incoming order. The order book is very liquid (*panel a*) when a great amount of orders is stored in each side of the order book and almost all quotes behind the best one are occupied. In such a situation a market order produces a small perturbation of the system and then a small price adjustment. On the contrary when a liquidity crisis occurs (*panel b*), the order book is characterized by few orders stored and by a large average gap between adjacent orders. In this case even a market order with a small volume can produce a dramatic price fluctuation of several ticks. In our discussion high and low liquidity situations are symmetric and refer to the structure of the order book

question is as crucial as the scientific investigation itself. The dynamics of financial markets is instead much closer to scientific phenomena than Social Sciences. The large amount of available data allows for systematic analysis, permits to test models and repeatability of observations. In addition as we have seen, the rules of functioning of order books are mechanical and therefore, even if order placement depends on the agents' willing, the framework of allowed actions is well-defined and limited. For these reasons, order books are a playground in which the methods of Statistical Physics have obtained academical and real (i.e. profitable) success due to the very defined framework provided by the double auction system. Before moving to statistical properties, in Fig. 5.3 we report how high frequency data usually appears.

5.2.1 Static Properties of Order Books

Overnight gap

Data are statistically homogeneous within the same day but the link between different days is intrinsically problematic. In fact, during the night, the price undergoes a large discontinuity which is of the same magnitude of the typical daily fluctuations (see Fig. 5.4). This phenomenon derives from the fact that during the closure of stock exchange, new important information can arrive and the opening auction can

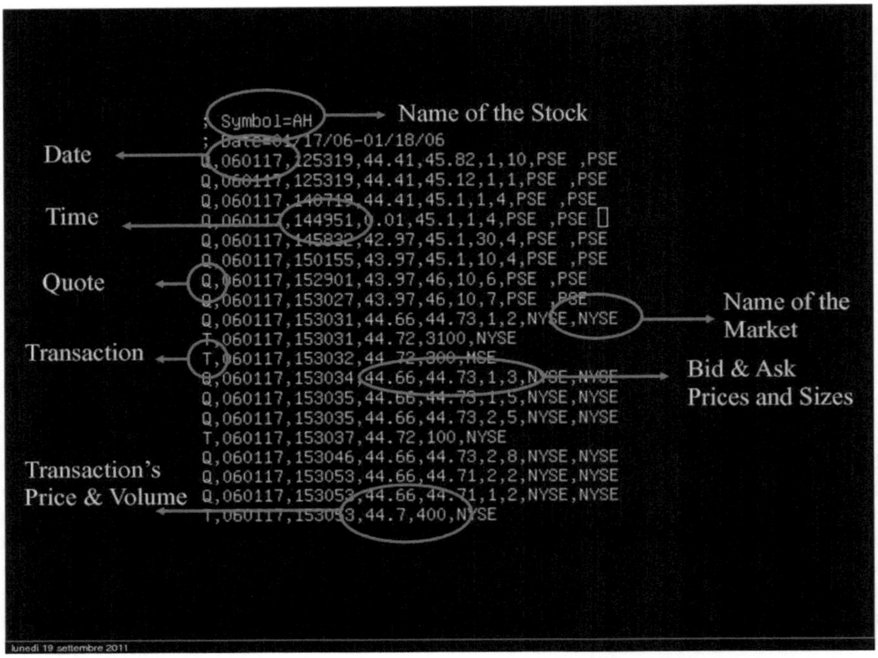

Fig. 5.3 The typical appearance of high frequency data. This particular example reports the high frequency data downloaded from NYSE

Fig. 5.4 Visual representation of the overnight gap, the night fluctuation has the same magnitude of the daily return

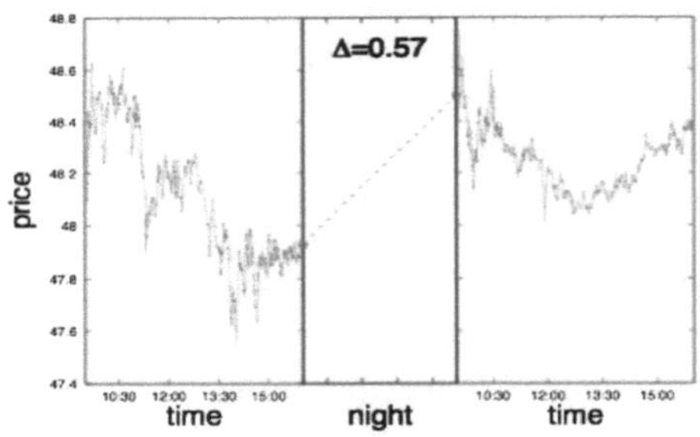

fix an opening price very far from the closure one or the same stock can be traded in different markets (i.e. Forex Market).

Order type and placement

Trades and orders can take place in two different ways. Off-book trades (or upstairs because they are discussed in private rooms above the crowded playground where brokers buy and sell stocks) are private transactions which usually occur between large institutions. The price of these transactions is not fixed by the public price and the details of the trade are make public only at the end of the operation. This kind of transactions accounts for 35 % of transactions and about 50 % of the traded volume.

On-book trades instead are the ones stored in the order book and these orders are all electronically recorded. This kind of orders is public, each trade can be seen by the other market participants. The analysis of the order book is usually focused on this kind of orders which are usually granted to academical institutions while the off-book transactions are kept secret.

One of the first question which can arise is related to the distribution of the limit orders. The distribution of the distance of the orders from the quote is found to be symmetric for ask or bid orders and it decays as a power law, in formula

$$P(\delta_p) \propto \frac{\delta_0^{\mu}}{(\delta_1 + \delta_p)^{1+\mu}} \tag{5.1}$$

where $\delta_p \geq 1$ is the distance of the limit order from the best quote expressed in ticks. The specific value of the exponent is market dependent, for instance at Paris Stock Exchange it is found to be $\mu = 0.6$ (see [1, 2]). This finding reveals a strategic placement of orders, in particular this is an evidence for a very rich ecology of strategy timescales, in fact the most distant orders (i.e. thousands of ticks) corresponds to long time strategies while orders, which are distance only few ticks, correspond to high frequency trading activity.

Order book profile

We have seen how far the orders are deposed by agents but we can also investigate how the order volumes are distributed with respect to the distance from the best quote. As in the previous case the ask side is found to be symmetric with respect to the bid one. The average profile is found to have a maximum some ticks away from the best quote even if the maximum flux of limit orders is observed at the best. This can be simply explained in terms of market orders which matches only orders at the best quote and of cancellation rate as we are going to see in the next section. It is worth noticing that even if the average profile is well-defined, this quantity does not give a good representation of a single realization of the order book because the volume fluctuations at each distance (and therefore the fluctuations with respect to the average profile) are well-described by a Gamma distribution with a very broad tail. In fact it is found that the ratio between the average volume at a certain distance and the magnitude of its typical fluctuations is about 1. In the end the average profile decays as a power law for orders which are very far from the best (>50 ticks) [1, 2].

Cancellation rate

The probability of a limit order cancellation is a decreasing function of the distance from the best [3, 4] and can be approximately described in terms of a power law. Therefore the maximum rate of cancellations is observed at the best price and this is one the reason for which the maximum of the average profile is not found at the best quote. This decreasing behavior is also interpreted as an effect due to the presence of automated trades.

5.2.2 Subtle Diffusive Behavior of Prices

In the first chapter we have said that one of the main Stylized Facts is the absence of simple arbitrage opportunity in financial markets. This correspond to the idea that the sign of the next price variation is unpredictable given the sign of the last one, that is the autocorrelation function of returns is identically equal to zero

$$\mathrm{E}[r_{t+\delta} r_t] = 0 \qquad \forall \delta > 0 \tag{5.2}$$

where we assume that the average return is zero without any loss of generality. We have already discussed that the stochastic process followed by prices is not a simple Random Walk. In this chapter we briefly describe the subtle way in which markets achieve the absence of simple correlation among returns. In fact when the autocorrelation function of the sign of transactions is investigated, differently from what expected by the evidence of the absence of simple arbitrage, we find that transaction signs are correlated and the correlation slowly decays as a power law [5–7]

$$\rho_\epsilon(\tau) \sim \mathrm{E}[\epsilon_t \epsilon_{t+\tau}] \sim \tau^{-\alpha} \tag{5.3}$$

where ϵ_t is the sign of the transaction (i.e. buy or sell order) and $\alpha \approx 0.6$ for the Paris Stock Exchange (see [6]). Then, given this evidence, one would expect a non zero correlation also in price returns since sell orders tend to produce downward movements while buy orders upward movements. Therefore in order to restore the absence of arbitrage there must exist non trivial correlations of the flux of limit orders which make the impact of this correlated signed orders almost null.

Let us now briefly explain the origin of this correlation. The sign correlation is due to fragmentation of large orders. In fact when large orders must be executed the financial institution which is performing this operation aims at minimizing the cost of the operation and this corresponds to minimize the impact of orders. By fragmenting the order in a large number of small orders which match the volume at the best price it can be shown that the impact of this procedure is smaller than the one produced by a single large order. This procedure is profitable because it allows for the revealing of the hidden liquidity of the order book. The hidden liquidity is roughly speaking the volume of the orders of the agents which are looking at the stock but they are waiting for a favorable moment or for the detection of some signals before sending their orders. In fact the real volumes available (that is agents who, at a certain time, would like to buy or to sell) is much larger than the one deriving from the limit order stored in the book.

The hidden liquidity is the key element that restores the unpredictability of the sign of the next return. In Fig. 5.5 we report a schematic representation of this mechanism. Traders detect the flux of sign-correlated orders, let us suppose sell orders (red boxes in the figure) which are, as said, the consequence of the splitting procedure of a large sell order. The agents which represent the hidden buy liquidity start to send orders of purchase at the best quote (or nearby), in such a way the buy liquidity (that is the

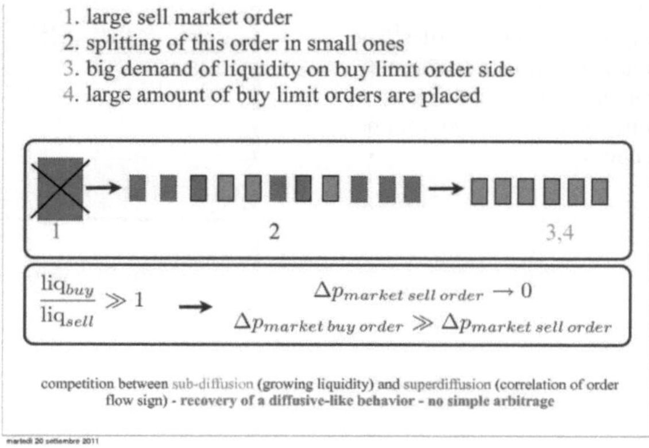

Fig. 5.5 Schematic representation of the mechanism which restores the absence of arbitrage

volume of buy limit orders) increases and the flux of correlated sell orders creates an unbalance of liquidity between the two sides of the order book. In fact buy liquidity is now much larger than sell liquidity since the hidden liquidity of the buy side has been revealed by the correlated sell orders. Therefore the impact of sell market orders become vanishing with respect to buy market orders and it can be shown that this competing effects restore the absence of arbitrage. For the mathematical details see [5–7].

5.2.3 Price Response to Orders

The study of the impact of orders as we have seen in the previous section is a crucial topic in order to minimize the cost of a transaction since the more the price is affected by the order sent by an agent, the more the price of the transactions will increase (for buy orders) and decrease (for sell order) and consequently get far from the optimal value at the beginning of the transaction.

The average response to incoming market orders is usually called price impact function and this function measures the average price variation due to an order of a given volume ω (Fig. 5.6). We can define two kinds of impact function, the virtual and the real one.

The virtual impact measures the average response in a static scenario: we freeze the order book at a certain time and we test the price response to virtual market orders with increasing volume, this procedure is repeated and the virtual impact function is the average of this virtual response to these virtual orders. This function is found to be concave with respect to the order volume but the explicit behavior depends on the stock market [1, 2, 8]. In fact in [1] the authors find that the impact is well described by $\ln \omega$, while in [8] by a power law ω^{α} with different exponent, $\alpha \approx 0.6$ and $\alpha \approx 0.3$. Concavity is due to the shape of the profile and to the response of the investors to a liquidity demand (liquidity provider).

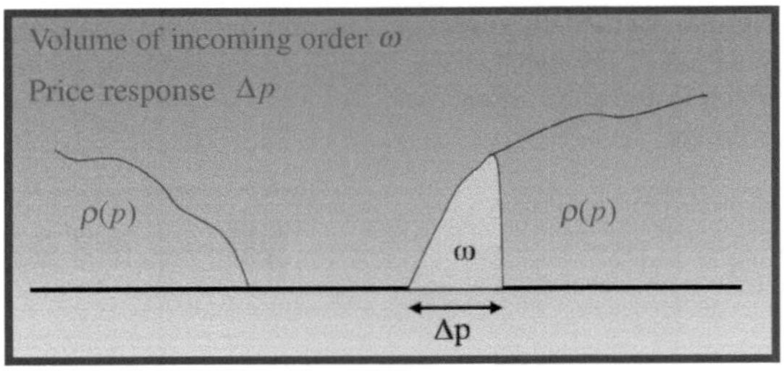

$$\omega = \int_{p(t)}^{p(t)+\Delta p} \rho(p) \mathrm{d}p$$

Fig. 5.6 Illustration of the impact of a market order. The figure also illustrates how to compute the virtual impact function for different volume values ω

The real impact function instead simply measures which is the real impact of real market orders. Given the fact that agents' actions are not independent on the flux of orders of the other market participants, we find that the real impact function is much smaller than virtual one. The real impact function is a very slowly increasing function with respect to order volume (sometimes it is even approximately constant). This can be explained by the fact that approximately 98 % of market orders have a volume lower or equal to the volume stored at the best quote [9, 10].

Gap distributions and returns

A different way to highlight the large discrepancy between virtual and real price impact functions consist in the investigation of how the gap of the order book are correlated with the price returns. In [10, 11] the distribution of price variations is compared with the distribution of the first gap, that is the distance between the best quote and the second best quote. The two distributions appear to be almost identical and this confirms that market order volumes are highly correlated with volumes at the best price and consequently they produce an almost constant price impact function differently from the virtual one. In [11] the authors also investigate a large number of stocks which are very different in their liquidity, capitalization, tick size, average traded volume, etc and they find that the relation between the exponent of the two distributions (gaps and price returns) is linear, confirming the correlation between market order volumes and volumes at the best quote.

References

1. Bouchaud, J.-P., Mezard, M., & Potters., M. (2002). Statistical properties of stock order books: empirical results and models. *Quantitative Finance*, *2*, 251.
2. Potters, M., & Bouchaud., J.-P. (2003). More statistical properties of order books and price impact. *Physica A*, *324*, 133.
3. Farmer, J., Patelli, P., & Zovko, I. (2005). The predictive power of zero intelligence in financial market. *PNAS*, *102*, 2254.
4. Mike, S., & Farmer, J. D. (2008). An empirical behavioral model of liquidity and volatility. *Journal of Economic Dynamics and Control*, *32*, 200.

5. Bouchaud, J.-P., Kockelkoren, J., & Potters, M. (2004). Random walks, liquidity molasses and critical response in financial markets. *Quantitative Finance*, *6*, 115.
6. Bouchaud, J.-P., Farmer, J. D., & Lillo, F. (2008). *How markets slowly digest changes in supply and demand*. Elsevier: Academic Press.
7. Farmer, J. D., Gerig, A., Lillo, F., & Mike., S. (2006s). Market efficiency and the long-memory of supply and demand: Is price impact variable and permanent or fixed and temporary. *Quantitative Finance*, *6*, 107.
8. Lillo, F., Farmer, J. D., & Mantegna, R. N. (2003). Master curve for price-impact function. *Nature*, *421*, 129.
9. Weber, P., & Rosenow, B. (2006). Large stock price changes: volume and liquidity. *Quantitative Finance*, *6*, 7.
10. Farmer, J. D., Gillemot, L., Lillo, F., Mike, S., & Sen., A. (2004). What really causes large price changes? *Quantitative Finance*, 4, 383.
11. Farmer, J. D., & Lillo., F. (2004) On the origin of power law tails in price fluctuations. *Quantitative Finance*, *4*, 7.

Chapter 6
Zero Intelligence Model for the Order Book Dynamics

6.1 Definition of the Model

In this chapter we study the price response function in presence of liquidity crisis from a theoretical point of view and to appropriately address this problem we introduce a model with a suitable microscopic dynamics. Our model can be defined as a *zero-intelligence* model as the ones introduced in [1, 2]. It should be noted that the specific questions that we consider cannot be addressed by the dynamics in [1, 2].

In fact in [1, 2] the deposition is modeled in term of a flow of orders driven by a Poisson process. The orders arrive at each quote at a certain fixed rate so that the deposition interval and the number of orders per unit of time are infinite but the orders per unit price interval are finite. The motivation of the authors of [1, 2] is that in such a way the order book can be never depleted and some finite size effects, due to a finite deposition interval, are avoided.

On the other hand in our model it is preferred an elementary *microscopic* mechanism for order deposition.

The difference between these two mechanisms for order deposition is the definition of the time unit. In the present chapter the time unit corresponds to the time (that is not fixed) between two following orders. Instead in [1, 2] the time unit corresponds to a physical and fixed amount of time (for example Δt equal to some minutes) during which several actions are made by investors. This permits a *coarse-grained* description in terms of average quantities, such as order flow.

For the specific implementation of our model we start by considering three mechanisms of deposition which will be compared with the empirical data.

We want to point out that only one of the three mechanisms gives rise to a realistic dynamics. Certain elements are however common to the three mechanisms and we start by discussing them.

In our set-up, as in a real order book, an order can be placed at a quote that is an integer multiple (positive or negative) of the tick size Δq. We do not investigate the effects of this parameter; therefore we set it equal to 1. For the sake of simplicity we assume that all orders (limit and market) have the same volume $\omega = 1$. The fact that

M. Cristelli, *Complexity in Financial Markets*,
Springer Theses, DOI: 10.1007/978-3-319-00723-6_6,

a market order has the same volume of a limit order is coherent with the observation that in more than 98–99 % of the cases a market order does not exceed the best limit order available [3].

In this model there is one mechanism for order creation, the limit orders deposition, and two for order annihilation, the arrival of market orders and the cancellation process of limit orders. The cancellation of a limit order occurs when the order has been stored in the book for τ time steps without being executed, where τ is a parameter to be fixed.[1] The balancing of the three mechanisms of creation/annihilation fixes the properties of the steady state that is reached very quickly (about 5×10^3 steps if τ ranges between 100 and 1000). Two limit orders can be also removed from the order book when the spread $s(t)$ is zero, that is when an incoming limit order is placed at the opposite best quote. However, these events are very unlikely (<1 % of the number of the market orders) and their effect is negligible.

We define the price as the mid-price $p(t) = [a(t) + b(t)]/2$ where $a(t)$ and $b(t)$ are the quotes of the best ask and of the best bid.

In our model we completely neglect the daily closures of the market and therefore the strong price variations during the night. The typical length of a run is 10^6 time steps (10^7 if a larger sample is needed) which corresponds to a real sample of about 5 years, estimating about 10^3 operations per day.

6.1.1 Order Deposition

In this section we give a detailed analysis of three possible cases for order deposition which we will test with respect to empirical data in order to select the most appropriate mechanism. Due to the symmetry of the order book, the probability that an order is a sell or a buy one is always the same.

Case 1

Once the nature (buy or sell) is determined, the order can be a market one with probability π and a limit one with probability $1 - \pi$, with $\pi < 0.5$. Limit orders will be placed in the interval $[b(t), b(t) + L]$ or $[a(t) - L, a(t)]$ depending on their nature (sell or buy respectively). The probability distribution within these intervals is considered uniform and L is a free parameter to be fixed. Hence the deposition interval is independent on the spread size.

[1] We know that this assumption, reasonable for the present study, is far to be realistic from the moment that analysis of real order books have shown that the lifetime of a limit order increases monotonically with its distance from the mid-price [4, 5].

Case 2

The order can be a market one with probability π and a limit one with probability $1 - \pi$ with $\pi < 0.5$ as in case 1. A limit order will be placed in the interval $[b(t) + 1, b(t) + k\,s(t)]$ or $[a(t) - k\,s(t), a(t) - 1]$ with $k > 1$ with a uniform distribution, being $s(t)$ the spread. In this case the length of the deposition interval depends on the spread size through a sort of auto-regressive process. This implies that the probability that a limit order produces a price variation (i.e. an order is placed inside the spread) is time-independent and approximately equal to k^{-1}. We anticipate that this second case will turn out to be the more realistic one with respect to the empirical data.

Case 3

The third case is inspired by the mechanism proposed in [5]. First a number ξ in the interval $[-L, L]$ is extracted. If $\xi < s(t)$ the order is a limit order otherwise it is a market one. If the order is a sell limit order its quote is $a(t) - \xi$. On the contrary if the order is a buy one its quote is $b(t) + \xi$. In such a way the probability of being a market or a limit order is not set a priori.

We have previously mentioned that our choice for order deposition produces a finite number of orders stored in the order book and a finite depth. This causes a finite size effect: the order book is subject to complete depletion with a cut off in those quantities linked with the book depth. However the probability of a complete depletion goes very quickly to 0 when the cancellation parameter τ is increased (this probability is less than 10^{-7} for $\tau > 200$).

Moreover we note that the uniform deposition mechanisms used in this chapter cannot reproduce in a realistic way all the empirical features of the order book; nevertheless this assumption has been made with the aim to define a minimal model still able to reproduce the main properties of the order book. Besides, the main goal of this work is the evaluation the effects of the granularity in price fluctuations, which depends on the static realization of the profile.

6.2 Calibration of Parameters and Preliminary Results

In this section we check the properties of the models we have introduced with respect to the empirical data. This analysis will also allow us to decide how to calibrate the parameters.

The empirical results in [3] indicate that a realistic value for π is approximately $1/3$. In case 3 the probability of being a market or a limit order is not fixed a priori. Nevertheless in this case we find that the fraction of market orders tends to a realistic value (about 0.32) when τ grows, and becomes substantially independent on τ when the probability of emptying the order book turns to be negligible (i.e. $\tau > 200$). This feature does not seem to depend crucially on the parameter L.

Table 6.1 Summary of the setting of the parameters

	Case 1	Case 2	Case 3
π	1/3	1/3	/
	$\tau < 200$ $n \ll 50$	$\tau < 300$ $n \ll 50$	$\tau < 200$ $n \ll 50$
τ	$200 < \tau < 500$ $50 < n < 100$	$300 < \tau < 750$ $50 < n < 100$	$200 < \tau < 500$ $50 < n < 100$
	$\tau > 500$ $n \gg 100$	$\tau > 750$ $n \gg 100$	$\tau > 500$ $n \gg 100$
L	200	/	100
k	/	3 or 4	/
General remarks	$\langle s \rangle \gg 20$	Realistic case $\langle s \rangle = 2.8$ for $k = 4$, $\tau = 400$	$\langle s \rangle \gg 20$

Only case 2 is able to reproduce reasonable values for the quantities of interest

We now have to fix the remaining parameters. Two quantities that can be calibrated on empirical data are the average spread and the average number of stored orders n. The parameters of the model which rule these quantities are τ, k (case 2) and L (case 1 and 3). Since it is not always possible to a find a configuration of all parameters able to reproduces these two quantities in a realistic way, we give priority to those sets of parameters that produce a reasonable average number of orders.

All three cases exhibit a realistic accumulation of orders (i.e. 50–100 orders per side) for τ ranging from 200 to 500. This clarifies the role of τ in driving the average number of orders in the steady state. In case 1 we choose $L = 200$ and in case 3 $L = 100$, in such a way the order book has approximately the same depth in both cases. For these two cases we applied the previous criterion of priority in order to have a realistic average number of orders. Case 2 appears to be very interesting because it exists a range of parameters for which we can reproduce both a realistic average spread and average number of orders. For instance for $\tau = 400$, $k = 4$ we obtain an average number of orders per side of about 60 and $\langle s \rangle \approx 2.8$ ticks. In Table 6.1 we report a summary of the results for the tested parameters.

It is worth noticing that the *auto-regressive* case 2 is also interesting with respect to the Stylized Facts. In fact only in this case we can observe some volatility clustering in the price increments (returns). All cases exhibit a return probability density function which decades as a power law (see Fig. 6.1) but only in the second case the exponent γ of the tail is compatible with the empirical observations ($-5 < \gamma < -2$). For instance in case 2 we find $\gamma = -4.1$ for $\tau = 200$ and $\gamma = -5.4$ for $\tau = 400$.

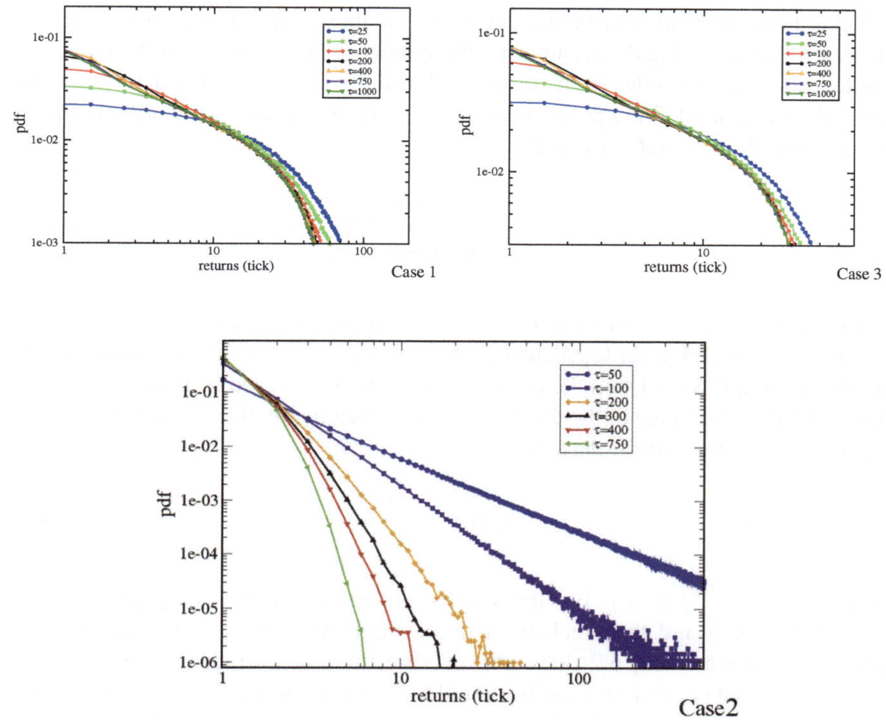

Fig. 6.1 Fat tails. We show the probability density function of returns for the three mechanisms for order deposition and for different values of τ. Because of the symmetry of the model the left tail is identical to the right one. We see that all three mechanisms give rise to a power law behavior but only the case 2 produces realistic values for the tail exponent γ: $\gamma = -3.1$ for $\tau = 200$ and $\gamma = -5.4$ for $\tau = 400$. In fact for cases 1 and 3 the exponent γ is larger than -2 which is not realistic. Moreover we note that the case 1 and 3 are very similar with respect to the tail exponent and to the cut off around 100 due to the deposition mechanism. We conclude that these two mechanisms produce the same results except for slight differences. Conversely the *auto-regressive* deposition mechanism of the case 2 appears to belong to a completely different class. This fact is also supported by the fact that the case 2 reproduces a certain volatility clustering differently from case 1 and 3

6.3 Operational Estimator of the Granularity

Given a certain configuration of the order book we want to define a way to measure the liquidity of it. In fact our objective is a detailed study of the relation between finite liquidity and price fluctuations.

We would like to describe the effect that the order book may not be compact but characterized by orders separated by voids. In order to perform this analysis we are going to introduce a function g which we call granularity. We propose a definition for granularity which is related to the inverse of the size of the average void between two adjacent quotes (gaps). At each time t the spread $s(t)$ sets a characteristic length which can be used as unit of measurement for these gaps. However we are also interested in measuring the granularity of the order book in a region far from the best quotes since the two sides of the order book usually have a depth much larger than $s(t)$. Therefore we define a partitioning of size $s(t)$ of one side of the order book and we perform the following average

$$\langle g(t)\rangle = \frac{1}{N(t)} \sum_{j=1}^{N(t)} \frac{n_{s(t),j}}{s(t)} \tag{6.1}$$

where $N(t)$ is the number of intervals of length $s(t)$ defined by the partition (we require $N(t) > 2$ in order to calculate $g(t)$) and $n_{s(t),j}$ is the volume of orders in an interval of length $s(t)$. From the definition of $N(t)$ we observe that $N(t)s(t) = \bar{L}$ is approximately the range of the order book and consequently the granularity g results equal to the average linear density of orders

$$\langle g(t)\rangle = \frac{\Omega}{\bar{L}} \tag{6.2}$$

where $\Omega = \sum_{j=1}^{N(t)} n_{s(t),j}$. For the sake of simplicity we now drop the temporal dependence of g and the brackets of the average. We note that the definition of granularity given in Eq. 6.2 does not measure the real average size of the voids because more than one order can be stored at the same quote. Therefore g^{-1} can be seen as the equivalent gap between two adjacent quotes in an hypothetical system where each quote can store only one order. We also observe that in the limit of a continuous order book, the price variation Δp produced by an order of volume ω would be

$$\int_0^{\Delta p} \rho(p)dp = \omega \tag{6.3}$$

where $\rho(p)$ is the density of the stored orders. If we approximate $\rho(p)$ with its average value g we find the following scaling relation for g and ω

$$\Delta p = \frac{\omega}{g}. \tag{6.4}$$

In this framework a liquidity crisis occurs when $g \to 0$, on the contrary the market is very liquid if $g \gg 1$. In fact when g is very large a great amount of orders can be executed without producing a significant price variation.

The granularity defined by Eq. 6.2 has the dimension of a $price^{-1}$. In order to obtain an absolute parameter for liquidity, we define the dimensionless liquidity as $\bar{g} = g\Delta q$. As we have seen in Sect. 7.1 at this stage we are not interested in the effect of the tick size Δq on the model so we have set it equal to 1. Consequently the dimensional and the dimensionless liquidity numerically coincide. Furthermore we want to point out that the definition of g given by Eq. 6.2 is not directly related to a

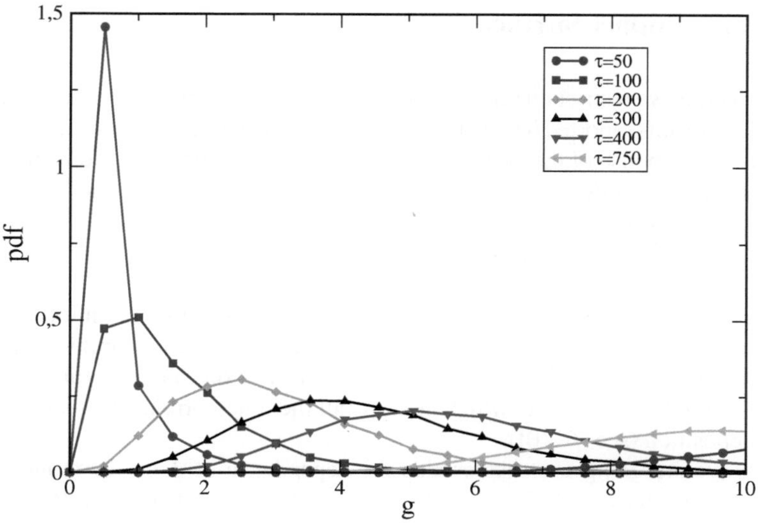

Fig. 6.2 Probability density function of the granularity g for the case 2 for different values of τ. Obviously large values of τ allow for a significant accumulation of orders at every quote (i.e. $g > 1$), despite this fact the probability of $g < 1$ and consequently of large price variation is not negligible for $\tau \approx 400$ which has been previously recognized as a suitable choice of this parameter

specific disposition in the order book but it gives only an average information about the void size.

In Fig. 6.2 we show the histogram of g for the model labelled as 'case 2'. The probability to observe a liquidity crisis, as expected, vanishes when $\tau \to \infty$ because of the large amount of orders stored in the order book. However, this probability cannot be neglected for values of τ ranging from 50 to 400 that we have previously recognized as the realistic ones.

It is important to note that in this chapter we give a different definition of granularity with respect to the definition found in [2]. In fact in that paper the authors define two dimensionless parameters to describe the discreteness of the system: the first, named granularity, is the dimensionless order size, the second one is the dimensionless tick size. In a certain sense both aspects are included in our definition of granularity and combined in a single parameter. These different choices are motivated by different purposes, while the authors of [2] gives the priority to the parameters that describe the order flows, differently we are going to focus our attention on how the discreteness influences the response function of the system.

6.4 Price Impact Surface

In this section we restrict our analysis to the 'case 2' mechanism for order deposition and all the results reported here correspond to $\tau = 400$ and $k = 4$. In order to analyze the average response of the system to an external perturbation (i.e. the volume of an incoming market order) after a certain time lag Δt, we introduce the Price Impact Function η defined as

$$\eta(\omega, \Delta t) = 2E[\Delta p(\Delta t)|\omega]. \quad (6.5)$$

One can consider two interesting limits of Eq. 6.5, the asymptotic limit ($\Delta t \to \infty$) and the instantaneous limit ($\Delta t \to 0$). In this chapter we restrict our investigation to the second case because the persistent effects on the market cannot be fully explained by a zero-intelligence model which neglects all the time-correlated structures of the order deposition (see [6–10]).

Now we are going to consider also the dependence of ϕ on the granularity g. In this way we have to deal with a Price Impact *Surface* rather than a Price Impact Function and we can analyze how the price response of the system depends both on the order volume and on the granularity g. Hence we define the Price Impact Surface as

$$\phi(\omega, g) = 2E[\Delta p|\omega, g]. \quad (6.6)$$

Before investigating the dependence of the price response surface on g we should verify the well-known empirical property that $\langle \phi(\omega, g)\rangle_g$ (i.e. the standard Price Impact Function defined as in Eq. 6.5) is a concave function with respect to the order volume ω. This is a manifestation of the fact that the order book is far from being in a linear regime. While the agreement on the concavity is almost universal, different functional form of the Impact Function have been given: the authors of [4] propose $\phi(\omega) \sim \ln \omega$, instead other authors propose $\phi(\omega) \sim \omega^\delta$ with exponent $\delta \approx 0.5$ [11] and $\delta \approx 0.3$ [6, 12]. In Fig. 6.3 we find that the surface averaged on g (black crosses) is concave and follows a power law

$$\langle \phi(\omega, g)\rangle_g \sim \omega^\delta \quad \text{with} \quad \delta \approx 0.59. \quad (6.7)$$

6.4.1 Price Impact Surface in the Direction of ω

Now we consider the Price Impact Surface as a function of ω for fixed values of g. We see from Fig. 6.3 that its behavior is a power law where the exponent β and the amplitude depend on g

$$\phi(\omega, g|g) \sim \omega^\beta. \quad (6.8)$$

This function is highly concave for small value of g and β grows for increasing values of g. However we notice that the dependence of the amplitude on g is far stronger

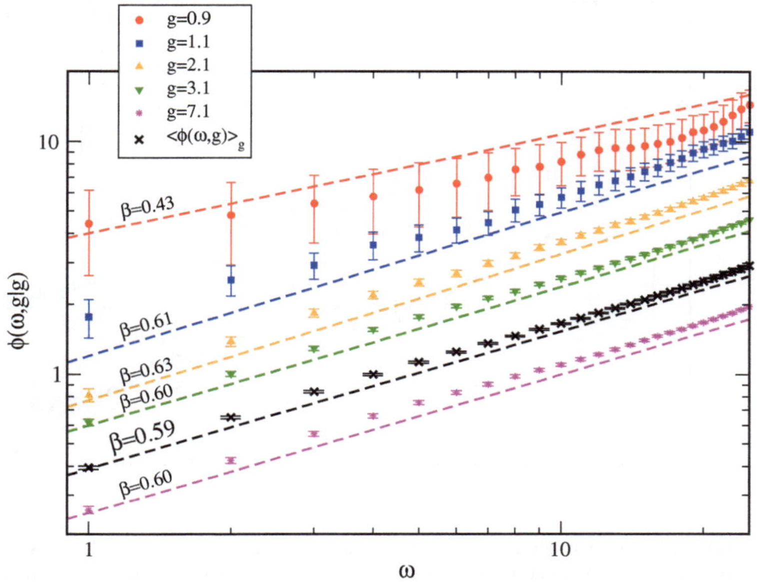

Fig. 6.3 Market Impact Surface in the direction of volumes ω for case 2 and $\tau = 400$. We can see that the function is concave as the one observed for real data. Its behavior is approximately a power law ω^{β}. The *dashed lines* correspond to the results of the fitting procedure and they are shifted for clarity purposes. The price response for large values of g cannot be larger than the one for small values of g. The *black crosses* represent the impact function averaged on g and correspond approximately to an exponent of 0.59

than the one of β. This implies, as expected, that the Impact Surface produces smaller variation of price in the limit of large g, even if the exponent β is larger.

It is worth noticing that the observed scaling behavior for ω is different from the one predicted by the zero order approximation made in Eq. 6.4. This is due to the peculiar shape of the order book profile whose maximum is peaked far from the best quote.

The results of Fig. 6.3 are qualitatively similar to those of Fig. 6 in [2] even if the different definitions of granularity do not permit a detailed comparison of the two models.

6.4.2 Price Impact Surface in the Direction of g

Now we analyze the price variation as a function of the granularity g for a fixed volume ω of the incoming market order. We find that the g-dependence of the surface is well represented by a power law

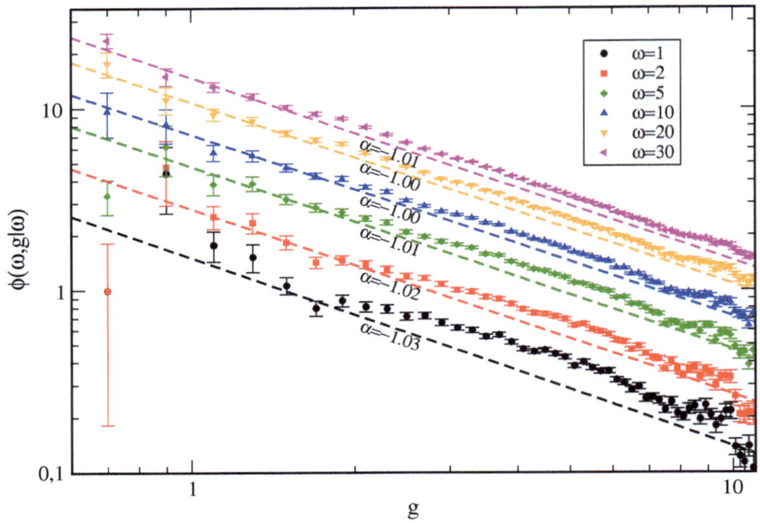

Fig. 6.4 Market Impact Surface in the direction of the liquidity g for case 2 and $\tau = 400$. In this direction the surface can be fitted with a power law g^{α} with a negative exponent. This implies that when g tends to 0 (i.e. when a liquidity crisis occurs) the price response can be very large even if the external perturbation (i.e. the volume ω of an incoming order) is very small. The exponent α is approximately independent on ω and its value is compatible with -1 within error bars

$$\phi(\omega, g|\omega) \sim g^{\alpha} \quad \text{with} \quad \alpha \approx -1. \tag{6.9}$$

In this case the exponent is negative and it appears to be nearly independent on ω (see Fig. 6.4). In fact we find that its value is -1 (within the error bars) for all values of ω. Consequently when the granularity vanishes the response of the system diverges. We can now quantitatively analyze how this limit is approached for finite (but large) values of g. For example if we consider an order of size $\omega = 10$, the average price variation induced when $g \approx 0.5$ is about ten times greater than the one observed when $g \gg 1$.

Therefore we have found that the amplification induced by the discreteness of the system is approximately proportional to the equivalent gap measured by g as predicted by the zero order approximation of Sect. 6.3. However we want to stress that for small values of g some deviations from the g^{-1} scaling behavior seem to be observed. This fact suggests that, when g vanishes, the deviation of the density $\rho(p)$ from the average density measured by g, is significant. Further investigation of this aspect will be considered in future works.

The fact that the exponent of the power law in the direction of g does not seem to exhibit a significant dependence on ω may suggest a possible factorization of the price impact surface, as we are going to verify in the next section.

6.4.3 Factorization of the Price Surface Impact

In [4] the authors observe that the empirical impact function admits the following factorization

$$\eta(\omega, \Delta t) = f(\omega)h(\Delta t). \tag{6.10}$$

Similarly we verify whether the Impact Surface Function could be expressed in the following way

$$\phi(\omega, g) = \varphi(\omega)\psi(g) \tag{6.11}$$

To check if the factorization of Eq. 6.11 is correct we have to verify if it exists a function of ω such that the impact surface divided by this function turns out to be independent on ω. This is equivalent to say that we would like to observe a complete collapse into a unique curve by rescaling with this function the surface ϕ. The simplest choice for this scaling function is the average impact function $\langle \phi(\omega, g) \rangle_g \sim \omega^{0.59}$. We report the rescaled functions for different values of ω in Fig. 6.5 and we observe a nearly perfect collapse for $\omega < 30$ within error bars.

In principle we cannot obtain a perfect factorization because of the dependence of the exponent β on g. Nevertheless we observe that this dependence is very weak, especially for $g > 1$ when $\beta \approx 0.60$. In fact in Fig. 6.5 one can appreciate some discrepancies only for $g < 1$. We can conclude that the factorization proposed in Eq. 6.11 approximately holds at least for small values of ω. We also expect that $\varphi(\omega, \Delta t \to 0)$ can be factorized with respect to its variables, instead the situation may be very complicated for the function $\psi(g, \Delta t \to 0)$ since g depend on time. We will consider these points in future investigations.

6.5 Summary and Perspectives

We have introduced a model of order book to study the generalized Price Impact Function (Price Impact Surface) and its dependence not only on order volume but also on granularity and we find that the granularity operates as a strong amplifier of price variations when a liquidity crisis takes place. Our result implies that the system response to an incoming order diverges in the limit of vanishing granularity. Furthermore agent-based models for financial markets usually do not take into account the problem of finite liquidity. In fact a common way to model the price movements in an agent based model is through the Walras' mechanism in which the price adjustments are proportional to the excess demand i.e. $\Delta p = \chi\, ED$. The coefficient χ and the excess demand ED are generally assumed to be independent on granularity. If, in first approximation, one interprets the Market Impact Function as the quantity the price movements are proportional to in the Walras' mechanism, then the previous coefficient χ must depends on granularity. Moreover we have observed that the Price Impact Surface can be factorized and we can try to identify the excess demand with

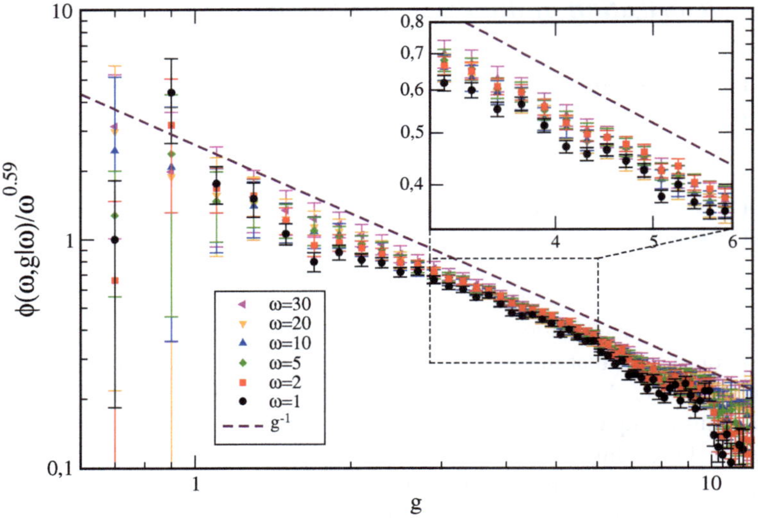

Fig. 6.5 Factorization of the Price Impact Surface. In order to verify if the Price Impact Surface admits the factorization $\phi(\omega, g) = \varphi(\omega)\psi(g)$, we rescale the surface in the direction of g (i.e. $\phi(\omega, g|\omega)$) with a function of ω. For the sake of simplicity we choose the Impact Function averaged on g that we have previously found to be proportional to $\omega^{0.59}$. We obtain a very good collapse of the rescaled functions within error bars. This confirms that the Price Impact Surface can be factorized at least for small values of ω. In the insert we show a magnification of the rescaled functions to highlight their statistical compatibility

$\varphi(\omega)$ and χ with $\psi(g)$. Consequently we aim at introducing this dependence on g in the *workable* agent based model we have introduced in previous papers [13–16]. In this framework we expect that even small unbalances of the market can produce large price adjustments if a liquidity crisis is present. Therefore the role of the amplification introduced by $\psi(g)$ could be one of the explanation of the breaking of the cause-effect relation and so that small perturbations can produce large fluctuations.

References

1. Daniels, M. G., Farmer, J. D., Gillemot, L., Iori, G., & Smith., E. (2003). Quantitative model of price diffusion and market friction based on trading as a mechanistic random process. *Physical Review Letters, 90*, 108102.
2. Smith, E., Farmer, J. D., Gillemot, L., & Krishnamurthy, S. (2003). Statistical theory of the continuous double auction. *Quantitative Finance, 3*, 481.
3. Farmer, J. D., Gillemot, L., Lillo, F., Mike, S, & Sen, A. (2004). What really causes large price changes? *Quantitative Finance, 4*, 383.
4. Potters, M. & Bouchaud., J.-P. (2003). More statistical properties of order books and price impact. *Physica A, 324*, 133.

5. Mike, S., & Farmer, J. D. (2008). An empirical behavioral model of liquidity and volatility. *Journal of Economic Dynamics and Control, 32*, 200.
6. Farmer, J. D., Gerig, A., Lillo, F., & Mike., S. (2006). Market efficiency and the long-memory of supply and demand: Is price impact variable and permanent or fixed and temporary. *Quantitative Finance, 6*, 107.
7. Bouchaud, J.-P., Kockelkoren, J., & Potters, M. (2004). Random walks, liquidity molasses and critical response in financial markets. *Quantitative Finance, 6*, 115.
8. Bouchaud, J.-P., Farmer, J. D., & Lillo, F. (2008). *How markets slowly digest changes in supply and demand*. Elsevier: Academic Press.
9. Bouchaud, J.-P., Gefen, Y., Potters, M., & Wyart., M. (2004). Fluctuations and response in financial markets: The subtle nature of random price changes. *Quantitative Finance, 4*, 176.
10. Lillo, F., Mike, S., & Farmer, J. D. (2005). Theory for long memory in supply and demand. *Physical Review E, 71*, 066122.
11. Plerou, V., Gopikrishnan, P., Gabaix, X., & Stanley., H. E. (2002). Quantifying stock-price response to demand fluctuations. *Physical Review E, 324*, 66.
12. Farmer, J. D. & Lillo., F. (2004). On the origin of power law tails in price fluctuations. *Quantitative Finance, 4*, 7.
13. Alfi, V., Pietronero, L., & Zaccaria, A. (2009). *Europhysics Letters, 86*, 58003.
14. Alfi, V., Cristelli, M., Pietronero, L., & Zaccaria, A. (2009). Minimal agent based model for financial markets. I: Origin and self-organization of stylized facts. *European Physical Journal B, 67*, 385. arXiv:0808.3562. http://arxiv.org/abs/0808.3562
15. Alfi, V., Cristelli, M., Pietronero, L., & Zaccaria, A. (2009). Minimal agent based model for financial markets. II: Statistical properties of the linear and multiplicative dynamics. *European Physical Journal B, 67*, 399. arXiv:0808.3565. http://arxiv.org/abs/0808.3565
16. Alfi, V., Cristelli, M., Pietronero, L., & Zaccaria, A. (2008). Mechanisms of self-organization and finite size effects in a minimal agents based model. *Journal of Statistics* (to be published). arXiv:0811.4256.

Chapter 7
Evidences of Strategic Placements of Orders

One of the most challenging issue of Social Sciences is the identification of new empirical features in order to discriminate and validate the various models proposed. We now investigate the effects induced by the discrete nature of order books. In fact, as we have already noticed, the price of the orders is not a continuous variable but can only be a multiple of a quantity called tick which is a fraction of the currency used. A first consequence of this aspect is the spontaneous emergence of asymmetries in the system. For instance, two configurations of the order book with an even spread or an odd one (in units of ticks) are not *a priori* equivalent for the mechanism of deposition of limit orders inside the spread. We investigate the fraction of odd spreads for a data set from the NYSE market finding indeed a strong evidence for this asymmetry. However, as we are going to see in detail in this chapter, the problem is more subtle than expected and also the agent strategies play an important role to explain quantitatively this phenomenon. The strategic order placement is also the origin of the non-trivial relaxation patterns observed when a spread fluctuation takes place.

7.1 Empirical Evidences

In our analysis of the order book we have considered a data set which spans a period of nearly 80 trading days between October 2004 and February 2005. This data set includes a series of high-frequency (tick-by-tick) information for 20 stocks from the NYSE market. These stocks have been chosen to be heterogeneous in their level of capitalization.[1] The information we have for each stock is the whole list of transactions and quotes, which are the prices of effective deals and of orders respectively. Our data set lacks the information about the whole order book apart of the best bid and the best ask. Therefore a new quote appears only if the best bid or the

[1] Market capitalization is defined as the number of shares of a company multiplied by their price. It is the simplest measure of a company's size.

M. Cristelli, *Complexity in Financial Markets*,
Springer Theses, DOI: 10.1007/978-3-319-00723-6_7,

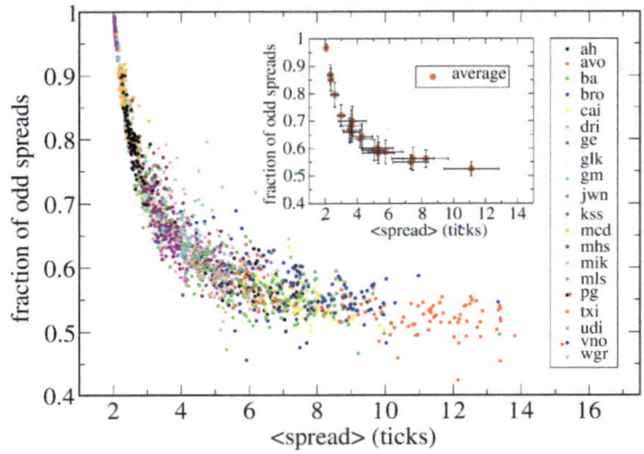

Fig. 7.1 Fraction of odd-valued spreads (in units of ticks) versus daily average spread for different stocks. We observe a systematic deviation from the symmetric case in which the fraction is 0.5. In the inset we plot a further average over a period of 80 days

best ask have been updated. From this data set we have reconstructed the sequence of market and limit orders by looking to the spread variations. If the spread has increased with respect to its previous value we refer to the event as a market order. On the contrary, the event is defined as a limit order if the spread has decreased. We assume that the probability that a cancellation of a limit order may change the spread is so small to be neglected. We have also expressed all the prices in units of ticks, so the spread results to be an integer number. At the end of this refinement, our data set is composed only by the series of the events in which the spread has changed, each labelled to be a market or a limit order. In this chapter we restrict our analysis to limit order events. To investigate the intrinsic asymmetry that a discrete and finite spread generates, we have firstly analyzed how the fraction of odd spreads depends on the average value of the spread for each stock. In Fig. 7.1 we have plotted, for each stock, the daily fraction of odd-valued spreads as a function of the average value of the spread. There are 80 different points for each stock and also a further average over all the days is plotted in the inset. We can observe that for almost all points the fraction of odd-valued spreads is larger than 0.5 which would be the expected value if the spread was very large ($s \to \infty$). This asymmetry is more marked for stocks with a smaller average spread and it goes diminishing while the average spread increases. A small average spread usually corresponds to stocks with a large capitalization. We are now going to investigate these results in the framework of the model we have introduced in a previous chapter and in [1].

7.2 A Model for Limit Order Deposition: Uniform Case

We propose a simple explanation of the evidences shown in Sect. 7.1 on the basis of the order book model introduced in [1] and in the previous chapter.

In such a framework we can evaluate a number of quantities, for example the probability that, given a spread s at time t, the new spread $s' \neq s$ at time $t + \Delta t$ is even or odd, where Δt is the time to wait to have a variation of the spread. The dependence on the parameter k is removed because we consider only the conditioned probability that an event occurs. In this chapter we indicate with s the value of the spread before an incoming event and with s' the new value of the spread consequent to the variation. In addition we restrict our analysis only to events due to limit order arrival. The probability to have an odd spread in the final state turns to be dependent on the parity of the spread s. Straightforward calculations give the probabilities

$$\begin{array}{ll} P(e|o,s) = \frac{1}{2} & P(o|e,s) = \frac{1}{2}\frac{s}{s-1} \\ P(o|o,s) = \frac{1}{2} & P(e|e,s) = \frac{1}{2}\frac{s-2}{(s-1)}. \end{array} \tag{7.1}$$

where o and e denote if the spread is odd or even respectively. One can see that $P(o|e) > P(e|e)$: given an even spread s in the initial state, the next spread s' has a larger probability to be odd rather than even. As expected, both $P(o|e)$ and $P(e|e)$ tend to 1/2 for $s \rightarrow \infty$. This simple argument gives a first explanation of the empirical evidence of an excess of odd spreads shown in the previous section. Now let us show that this simple observation is not able to take into account all the asymmetric pattern observed in Fig. 7.1

7.3 Data Analysis

The dispersion of the points plotted in Fig. 7.1 can be traced back to the spurious effect introduced by the fact that many different spreads s give their contribution to the final average. Therefore the next step is to investigate the frequency of odd spreads conditioned to a given s, in order to compare Eq. 7.1 to real data.

We can identify the variations of the spread caused by limit orders imposing the condition $s > s'$ because only limit orders inside the spread can decrease its value (the same argument has been also followed, for example, by [2]). In Fig. 7.2 we show the conditional probability to have an odd spread s' starting from a spread s, as a function of s. The pattern strongly oscillates around the value 1/2: an even spread is most likely followed by an odd one as predicted by Eq. 7.1, but surprisingly also the vice-versa is true.

This apparently strange behavior can be attributed to a non-uniform depositions of limit orders inside the spread. We studied the distribution of the spread variations conditioned to a given value of the spread s for real data. We found that a consistent fraction $\alpha \sim 0.7$ of the limit orders inside the spread is placed at the quote adjacent to the best, as we show in Fig. 7.3. A reasonable way to model this non-uniform distribution of the limit orders inside the spread is through a piecewise constant function. In this way the probability to put a limit order at the first adjacent quote is α, and the probability to put the order in one of the other remaining quotes is equally

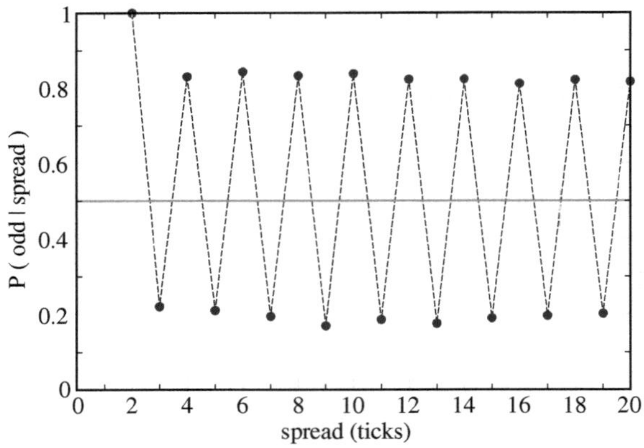

Fig. 7.2 Conditional probability to have an odd-valued spread s' given an initial spread s for real data. The plot shows clearly that with high probability an odd spread is followed by an even one (0.8) and vice-versa

distributed. This tendency can be also interpreted in terms of agents' strategies. In fact, the placement an order far from the (previous) best bid or ask can be seen as a risky operation in which the agent, by disagreeing with other agents' evaluations, tries to trade quickly paying a kind of virtual cost equal to the distance between the best and her order quote. In other words, an order near the best is the most conservative position able to change the spread. The next step is to find how much the probability α to place a limit order near the corresponding best depends on the spread s and on the stock considered. In Fig. 7.4 we plot α as a function of s for three stocks which cover a wide range of capitalization, finding very different behaviors depending on

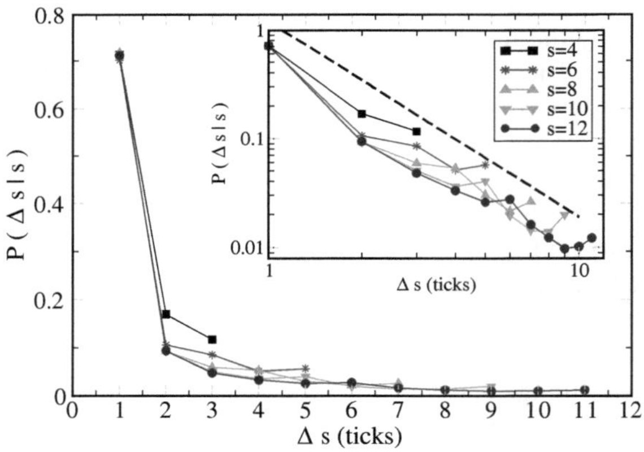

Fig. 7.3 Probability to have a variation Δs of the spread given an initial s for different spread values (empirical data). The most probable variation is always $\Delta s = 1$. The probabilities of the other variations ($\Delta s = 2, 3, \ldots$) are weakly decreasing functions and, in first approximation, can be considered as constant. The *dashed line* is the power law with exponent 1.8 found in [3] for London Stock Exchange. In both cases $\Delta s = 1$ is highly preferred

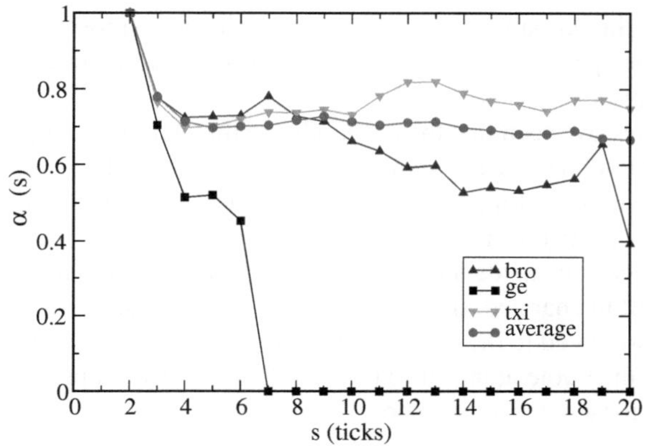

Fig. 7.4 Probability $\alpha(s)$ to place an order at the quote adjacent to the best one as a function of the spread. We have plotted this probabilities for three representative stocks and also an average over all the 20 stocks. The very different values observed can be explained by considering the different statistics of the stocks. To properly address this effect we have performed an average over the 20 stocks of our data set. The result is a value of $\alpha(s)$ approximately constant

the stocks and on the values of the spreads. One possible explanation for this variety of behaviors is the lack of statistics: in fact, liquid stocks (that is, stocks characterized by fast order execution and small transaction impact on the price) usually have small spreads, and hence the statistics for large spreads is poor, and vice-versa for illiquid stocks. We can appropriately weight the contribution of different stocks by averaging on all the data. The resulting curve is approximately a constant as a function of the spread.

7.4 A Model for Limit Order Deposition: Non-uniform Case

In the previous section we have observed a systematic deviation of the experimental probabilities $P(o|e,s)$ and $P(e|o,s)$ from the ones of Eq. 7.1 derived from the hypothesis of uniform order deposition. In this section we are going to show that this discrepancy is mainly due to the non-uniform probability of order placement inside the spread, and therefore to agents' strategies.

In order to include the effect of a non-uniform order deposition, we can generalize Eq. 7.1 in the following way

$$P(e|o,s) = \sum_{j=1}^{\frac{s-1}{2}} g(2j|s) \qquad P(o|e,s) = \sum_{j=0}^{\frac{s-2}{2}} g(2j+1|s)$$
$$P(o|o,s) = \sum_{j=0}^{\frac{s-3}{2}} g(2j+1|s) \quad P(e|e,s) = \sum_{j=1}^{\frac{s-2}{2}} g(2j|s). \tag{7.2}$$

where $g(i|s)$ is the probability mass function of the deposition mechanism for limit orders inside the spread and $i = 1, \ldots, s-1$ is the distance from the best quote of the placement price.

In Fig. 7.4 we have seen that the probability α that a limit order produces a spread variation equal to one is weakly dependent on the value of the spread. This suggests a simple approximation for $g(i|s)$ with a piecewise function. Now we discuss how to introduce this property in our model.

If a limit order falls inside the book or at the best quotes the mechanism for order deposition is left unchanged. Instead, if a limit order is placed inside the spread, the probabilities associated to the $s-1$ available quotes are no more uniform but highly peaked around the quote adjacent to the best. The deposition probabilities for a buy or a sell order, for a given s, become

$$g(i|s) = \begin{cases} g(1|s) = \alpha & \\ g(i|s) = \frac{1-\alpha}{s-2} & i = 2, ..., s-1. \end{cases} \tag{7.3}$$

For buy orders the index i is the distance from the best bid while for sell orders the index i is the distance from the best ask. In the previous section we have seen that in real markets α can be approximately considered as constant and its value is about 0.7. Clearly Eq. 7.3 is meaningful only for $s \geq 3$ since for $s = 1$ limit orders cannot fall inside the spread and for $s = 2$ a limit order inside the spread is always placed at the quote adjacent to the best one.

It follows directly from Eq. 7.3 that the transition probabilities now read

$$\begin{array}{ll} P(e|o,s) = \alpha + \frac{1-\alpha}{2}\frac{s-3}{s-2} & P(o|e,s) = \alpha + \frac{1-\alpha}{2} \\ P(o|o,s) = \frac{1-\alpha}{2}\frac{s-1}{s-2} & P(e|e,s) = \frac{1-\alpha}{2}. \end{array} \tag{7.4}$$

In Fig. 7.5 we plot a comparison between the expression of Eq. 7.4 and the corresponding probabilities evaluated from our data set. The experimental results are obtained by averaging over all the 80 trading days and over all the 20 stocks. The oscillating behavior can be explained by considering that a variation of spread of one tick ($\Delta s = 1$) is highly preferred with respect to other variations. Hence an odd spread goes more likely to an even one and vice-versa. We are now able to separate the two effects that contribute to enhance the fraction of odd spreads and produce the pattern of Fig. 7.1 through a simple Monte Carlo simulation. These two contributions are the intrinsic asymmetry due to discreteness and non-uniformity of order deposition. As initial conditions we generate some sequences of spreads with different means, in order to represent different virtual stocks. Starting from each sequence we simulate the transition to odd or even spreads s' according to the probabilities of Eqs. 7.1 and 7.4. In such a way we can evaluate the average fraction of odd spreads for each virtual stock. In Fig. 7.6 we compare the empirical average asymmetry with the results of the Monte Carlo simulations in the two cases just mentioned. The intrinsic asymmetry alone is not able to fit properly the experimental data which instead are well reproduced by considering the two combined effects. Nevertheless

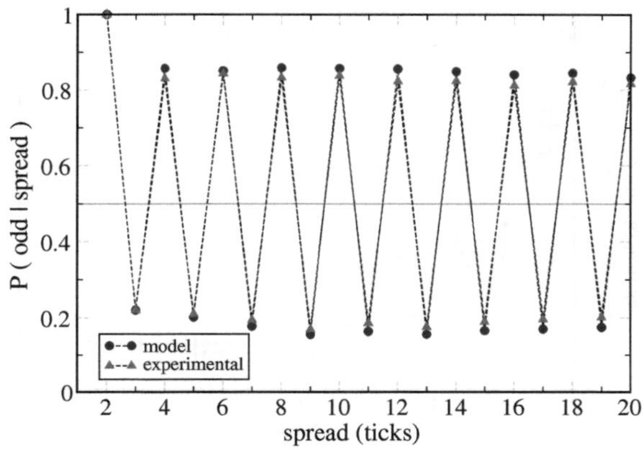

Fig. 7.5 Comparison between the experimental data and the phenomenological model described in the text. The phenomenological probabilities of the model (*dots*) show a small systematic overestimation of the oscillations with respect to the experimental ones (*triangles*). This effect can be easily understood in terms of the approximation considered in Eq. 7.3

we can observe some deviations for large spreads (> 6). This is due to the fact that we have assumed a constant probability to place order at quotes different from the one adjacent to the best. When the spread grows the error introduced by this assumption becomes larger. It is worth noticing that here we are neglecting the correlations between the spread values. Anyway we can recover the experimental behavior even with this uncorrelated sequence since we are averaging on times far longer than the time scales of the spread correlation.

7.5 The Effects of the Strategic Deposition of Orders

An interesting question concerns the role of the parameter α and how the non-uniform deposition inside the spread affects the order book statistical properties. The model allows this kind of investigation and permits to study the effect of different strategies of order placement inside the spread.

A numerical simulation reveals that if the same set of parameters of the previous chapter (and see [1]) is used, the spread dynamics diverges for $\alpha > 0.85$. To explain this effect we have to consider the interplay between market and limit orders. Market orders tend to move away the two best quotes eroding the book. Instead limit orders tend to reduce the spread by coupling the processes followed by best ask and best bid. In such a way the process for the spread is somehow stationary. In this framework the deposition rules play an important role in softening or strengthening the coupling action of limit orders. In fact the coupling between $a(t)$ and $b(t)$ is ruled by two elements. The first one is the fraction of orders which fall inside the spread. A larger fraction of these orders produces a stronger coupling between the best quotes. The

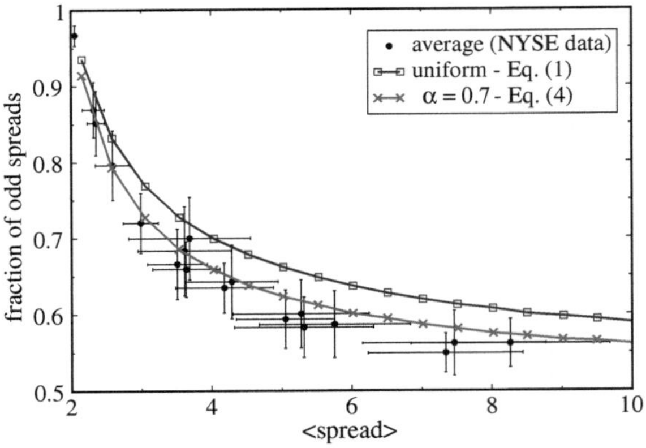

Fig. 7.6 Fraction of odd-valued spread versus average spread for experimental data and for two different Monte Carlo simulations. The first simulation (*squares*) is performed using the uniform order deposition mechanism. The second one (*crosses*) is instead performed with the non uniform mechanism. These two simulations permit to investigate the two contributions to the asymmetry between odd and even spreads. As expected we observe that the intrinsic asymmetry does not reproduce the experimental pattern (*dots*) while the two combined effects fit the experimental data very well. The small discrepancy found for large spreads is originated by the approximation made in Eq. 7.3

second is the mechanism of deposition of orders inside the spread, which governs the average spread variation produced by a limit order. In order to analyze the effect of the deposition mechanism we can reason as it follows. By considering the symmetry of the uniform case, we find

$$\frac{\langle \Delta s \rangle}{s} = \frac{1}{2} \tag{7.5}$$

while in the non-uniform case, from Eq. 7.4, we obtain

$$\frac{\langle \Delta s \rangle}{s} = \frac{\alpha}{s} + \frac{(1-\alpha)[s(s-1)-2]}{2s(s-2)}. \tag{7.6}$$

We recall that $s \geq 3$ since for $s = 1$ limit orders cannot be placed inside the spread and for $s = 2$ we always have $\Delta s/s = 1/2$. The inequality $\langle \Delta s \rangle/s \geq 1/2$ for Eq. 7.6 is satisfied when

$$2 \leq s \leq \frac{\alpha+1}{\alpha} \tag{7.7}$$

and we observe that $(\alpha + 1)/\alpha > 3$ only for $\alpha < 0.5$ (see Fig. 7.7). Consequently the average spread variation and the coupling action of limit orders in the non-uniform case are never larger than the one produced by the uniform mechanism in the region of parameters investigated ($s \geq 3$ and $\alpha > 0.5$).

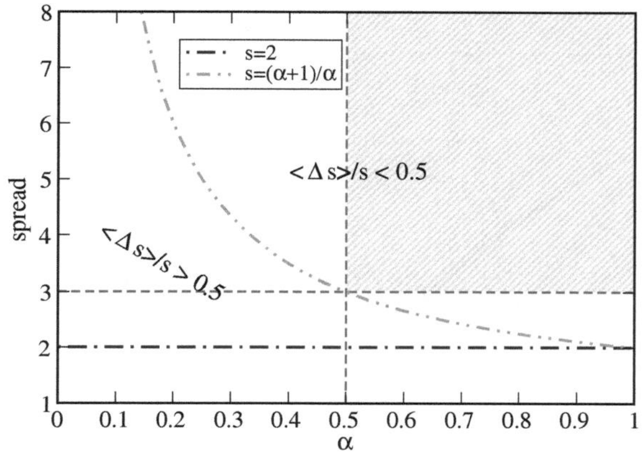

Fig. 7.7 Phase diagram of the relative average spread variation for the non-uniform mechanism. The relative average spread variation $\langle \Delta s \rangle / s$ is 0.5 for the uniform case. $\langle \Delta s \rangle / s$ is larger than 0.5 only for spreads included in the region between $s = 2$ and $s = (\alpha + 1)/\alpha$. The highlighted region ($s \geq 3$ and $\alpha > 0.5$) corresponds to a realistic scenario. We see that in this region the relative average variation of the spread produced by a limit order is always smaller than 0.5. Therefore the non-uniform deposition reduces the coupling action of limit orders with respect to the uniform case

Fixed $\alpha \leq 0.8$ we can analyze the properties of our model and we will come back on the problem of the stability with respect to α at the end of this section. The set of figures in Fig. 7.8 clarifies the role of a non-uniform deposition inside the spread.

In Fig. 7.8a, b we have plotted the probability density functions for the price variations (returns) and for the spreads respectively. We observe that in the non-uniform case the system produces larger fluctuations and larger average spreads, as we expected from the previous discussion. It is interesting to notice that when the system tends to a regime in which the order book is always compact, i.e. in which most of the quotes inside the book are occupied, the statistical properties becomes nearly independent on the deposition details ($\pi \leq 0.25$).

Figure 7.8c reveals that the non-uniform deposition also amplifies the fluctuations of the granularity g, defined as the linear density of the volume stored in a side of the order book [1]. In particular the non-uniform deposition shows a non-trivial temporal structure for granularity that resembles an intermittency phenomenon. Since most of the arriving limit orders are placed adjacent to the best quote the order book stays for long times in a quiet and compact state characterized by an average spread whose value is nearly one. This is the dominant regime of our simulated order book, but sometimes bursts of volatility are observed. In fact, when a large fluctuation of spread occurs, the autoregressive property and the non-uniformity of the limit order deposition make the relaxation towards the compact state very slow. This intermittency is directly related to the volatility correlation that is far longer in the non-uniform case than in the uniform one (see Fig. 7.8d). We want to stress that a small volatility clustering is already present in the uniform case. Its origin can be

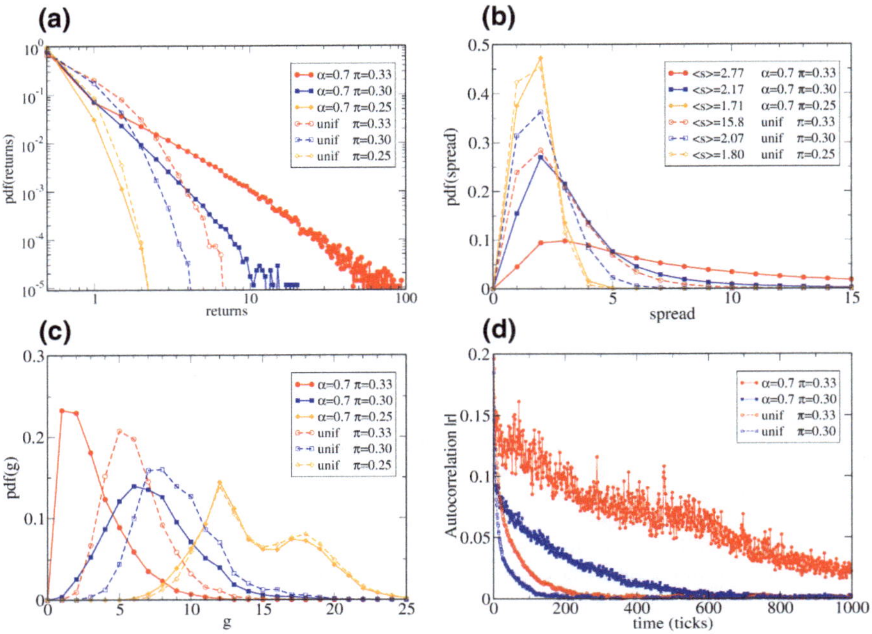

Fig. 7.8 Statistical properties of the simulated order book for the uniform and non-uniform case. In panels **a** and **b** we plot the probability density functions for the returns (Δp) and spreads respectively for different market order rates (π). These plots show that the average fluctuations of spreads and returns are larger in the non-uniform case. When the order book turns to a compact regime ($\pi \leq 0.25$), the statistical properties of the model become nearly independent on the deposition details. Panel (**c**) reveals that the non-uniform deposition produces non-trivial fluctuations of liquidity/granularity. In panel **d** we plot the autocorrelation of the absolute values of returns and we observe the presence of persistent volatility. The decay of the autocorrelation function of the absolute values of returns is exponential except for short time lags where spurious effects take place. This persistent behavior suggests the presence of an intermittent dynamics for the order book characterized by bursts of volatility

traced back to the dependence on the past spread values of the deposition mechanism. This simple effect introduces an exponential correlation and so fixes a characteristic time-scale. The non-uniform deposition instead amplifies this effect because the mechanism sets a further and longer time-scale that depends on α (the correlation length increases for increasing values of α). The correlation functions in Fig. 7.8d obey to an exponential decay except for short time lags where some spurious effects take place. The bursts of volatility of the non-uniform case are even more evident if we represent the complete order book. In Fig. 7.9, that corresponds to the uniform case, the order book is always compact. Instead in Fig. 7.10 we plot the non-uniform case and we find that the system stays for most of the time in a regime which is very similar to the one of Fig. 7.9 (regions I, III, V) but sometimes regions characterized by large spreads and large price movements appear (regions II, IV, VI). It is worth

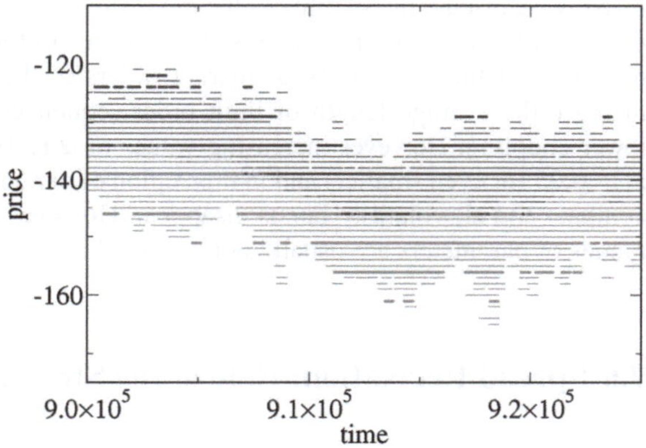

Fig. 7.9 Snapshot of the simulated order book in the uniform case ($\pi = 0.3$). The order book is always in a compact regime in which the average spread is nearly 1 and a small and local volatility clustering is observed due to the autoregressive deposition rules. For higher values of π small deviations from the compact regime appear but these phenomena cannot be compared to the intermittency produced when the order deposition is non-uniform

noticing that in this model large price fluctuations emerge spontaneously, being triggered by a random spread variation (and vice-versa). This mechanism resembles the phenomenon of self-organized criticality [4, 5].

It can be argued that also for larger values of π the uniform case could produce this intermittency because this correspond to an increase of the time-scale on which the autoregressive mechanism is able to produce local volatility clustering as we can observe in Fig. 7.8d. Nevertheless the correlation is still far shorter when $\pi = 0.33$ with respect to the one generated by the non-uniform case with $\pi = 0.3$. Furthermore the magnitude of the correlation is smaller than the one of the non-uniform case with lower values of π. Finally a visual inspection of the order book reveals that an intermittency of a kind appears but it is very small and we never obtain an order book with sudden transitions from a compact regime to a volatile regime as it happens in region II of Fig. 7.10.

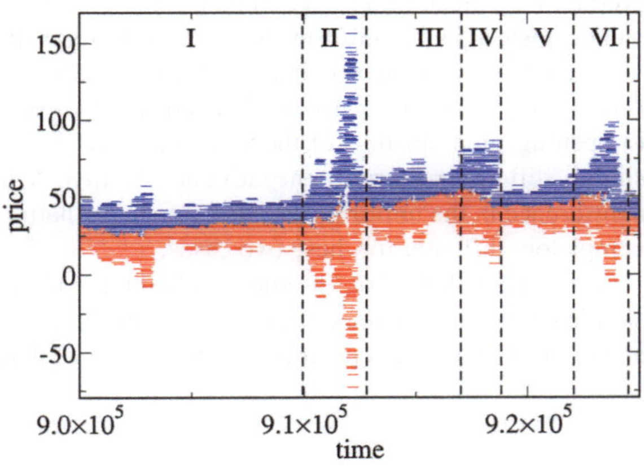

Fig. 7.10 Snapshot of the simulated order book in the non-uniform case ($\pi = 0.3$ and $\alpha = 0.7$). These deposition rules produce an order book which is typically quiet and compact as in the uniform case (region I, III, V) but exhibits bursts of activity due to large fluctuations of the spread (region II, IV, VI). The system gives rise to a sort of intermittency since volatility is very persistent and clustered

Now we discuss the stability of the model for $\alpha \geq 0.8$. We have seen in Fig. 7.8a, b that, in order to make the system stable with respect to α, a possible solution is the reduction of the probability of market orders π. In such a way it is possible to increase the average length of limit order sequences and then to compensate the lower coupling. However increasing values of α (>0.8) would imply a choice for $\pi \approx 0.25$ (or even smaller) and in this region of parameters the order book is always compact and all volatility bursts disappear. However it is worth noticing that these ranges of parameters are usually not observed in empirical data.

7.6 Spread Relaxation: Role of the Strategic Order Placement

Ponzi et al. in [3] studied the relaxation dynamics after an opening or a closing of the spread in LSE[2] order book. They find a slow relaxation of the spread towards the mean value. This decay is compatible with a power law and the authors argue that the absence of a characteristic time scale is due to the presence of a strategic placement of the orders.

We want to verify this empirical findings in the framework of the model introduced in the previous sections. In this respect we define, as in [3], the quantity

$$G(\tau|\Delta) = E[s(t+\tau)|s(t) - s(t-1) = \Delta] - E[s(t)] \tag{7.8}$$

where $E[\cdot]$ is the average on the whole time series, Δ is the spread variation occurred at time t (i.e. $\tau = 0$) and $\tau \geq 0$. It is worth noticing that, in our model, τ is expressed in time event units differently from the analysis in [3] which is performed in physical time. The mapping between these two time units is not necessarily linear and therefore a quantitative agreement should not be always expected.

We perform the spread relaxation analysis in both the uniform case and non-uniform case and we report the results in Fig. 7.11. The non-uniform mechanism produces a plateau of a kind and then for $\tau > 100$ a faster decay to *normal* spread values. Instead in the uniform case the spread relaxation is much faster than the previous case. Now we analyze in detail the non-uniform case. In Fig. 7.12 we report the function $G(\tau|\Delta)$ for positive and negative values of Δ (corresponding respectively to openings and closings of the spread at time $\tau = 0$). As in [3], we observe two slightly different patterns for negative and positive Δ. Instead, we do not observe such a difference in the uniform case, this means the nature of the relaxation dynamics is completely different for these two cases.

We can conclude that, in order to obtain a realistic spread relaxation function, a simple change to a pure zero-intelligence model consists in the non-uniform mechanism of limit orders deposition inside the spread described in Sect. 7.4.

[2] London Stock Exchange.

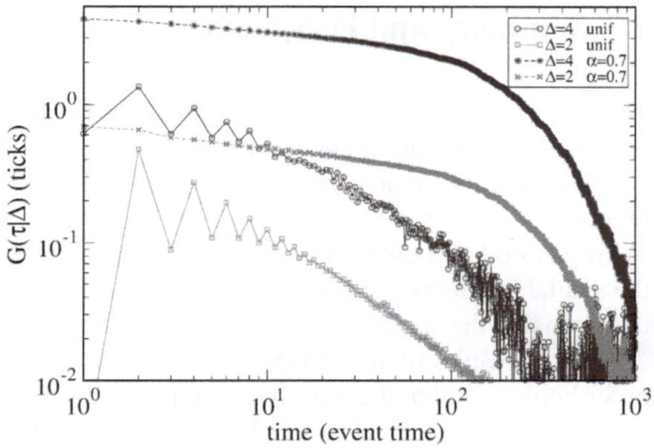

Fig. 7.11 Spread decay given Δ in the uniform case (*solid lines*) and in the non-uniform case (*dashed lines*). We observe a much slower decay in the latter case

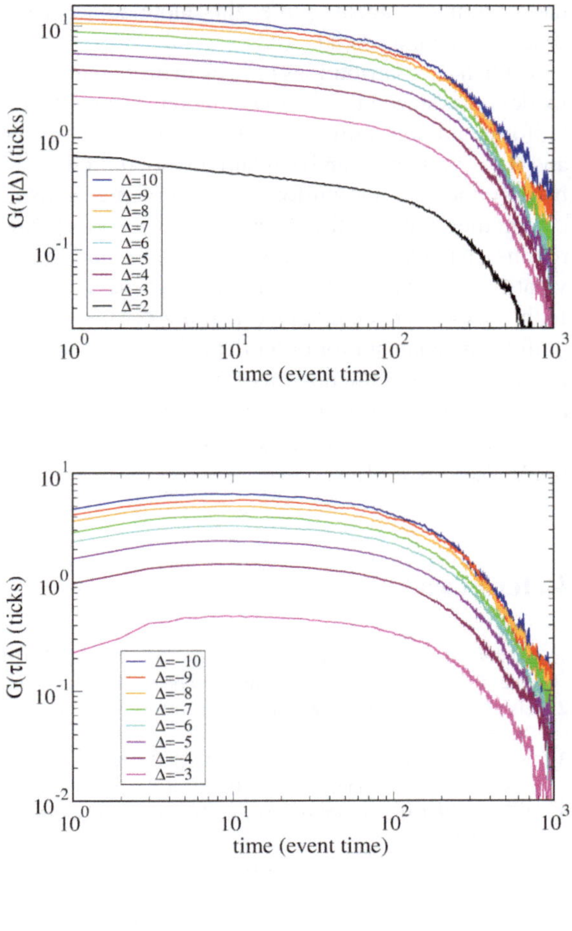

Fig. 7.12 Spread decay in the non-uniform case for positive (*top panel*) and negative (*bottom panel*) values of Δ. The value of $G(\tau|\Delta)$ is higher for higher absolute values of Δ. The patterns found in our model are very similar to the empirical results of Ponzi et al. in [3]

7.7 Summary and Perspectives

The order book is a system which is intrinsically discrete, for instance the quotes of placement of an order must be a multiple of the tick size. We find that the effects of this discreteness are non-trivial and produce deviations with respect to a continuous regime.

In fact odd and even spreads are not equivalent for limit order deposition when the available quotes inside the spread are discrete. Indeed if a uniform deposition of orders inside the spread is assumed, the system spontaneously prefers odd spreads.

One of results of this chapter confirms that this asymmetry is present in real order books and that the fraction of odd spreads is significantly above 0.5. However the asymmetry observed cannot be explained quantitatively only by considering the discrete nature of the order book. In fact a second effect also contributes to modulate the asymmetry: agents prefer the quote adjacent to the best one when they place orders inside the spread.

Both these contributions have been investigated in the framework of a microscopic model introduced in a previous chapter. The model permits to compare the effects of uniform and non-uniform deposition mechanisms for limit orders inside the spread and the asymmetry can be quantitatively reproduced in the framework of our model by introducing a non-uniform deposition mechanism.

Another result is the emergence of a sort of intermittent dynamics in which a regime characterized by a compact and liquid order book dominates but bursts of volatility also appear. This intermittent behavior is also observed in real order books. In this respect, in the framework of our model, we compare the uniform and the non-uniform mechanism for order deposition with respect to the dynamics of the spread relaxation when a fluctuation occurs. We find that the introduction of a simple rule of order placement is sufficient to reproduce the peculiar pattern observed in real data (see [3]). The observed relaxation cannot be explained by a pure zero-intelligence model (see also [6]).

References

1. Cristelli, M., Alfi, V., Pietronero, L., & Zaccaria, A. (2010). Liquidity crisis, granularity of the order book and price fluctuations. *European Physical Journal B*, *73*, 41.
2. Eisler, Z., Kértesz, J., & Lillo, F. (2007). The limit order book on different time scales. *PROC.SPIE*, *6601*, 0G.
3. Ponzi, A., Lillo, F., & Mantegna, R. (2009). Market reaction to a bid-ask spread change: A power-law relaxation dynamics. *Physical Review E*, *80*, 016112.
4. Bak, P. (1996). *How nature works: the science of self-organised criticality*. New York: Copernicus Press.
5. Jensen, H. (1998). *Self-organized criticality*. Cambridge: Cambridge University Press.
6. Toth, B., Kértesz, J., & Farmer, J. (2009). Studies of the limit order book around large price changes. *European Physical Journal B*, *71*, 449.

Chapter 8
Quantitative Analysis of Technical Trading

Right to now we have seen that Physics and mathematical methods and concepts derived from Complex System Theory and Statistical Physics can give a new insight to economic and financial problem. As briefly discussed in Chap. 3 chartist strategies, i.e. strategies based on the analysis of trends and recurrent patterns, should not exist according to the classical theory of financial markets because prices should follow their fundamental values. It is instead well-known that chartists exist and operates on different time scales. In this chapter we investigate if a specific chartist strategy, that is if there exist special values on which prices tend to bounce, produces a measurable effect on the statistical properties of the price series. As we are going to see we preliminarily must formalize this strategy in a mathematical framework in order to quantify its effect on the price time series under investigation.

8.1 Technical Analysis

8.1.1 The Classical and the Technical Approaches

The classical approach in the study of the market dynamics is to build a stochastic model for the price dynamics with the so called martingale property [1]

$$E(x_{t+1}|x_t, x_{t-1}, \ldots, x_0) = x_t \quad \forall t. \tag{8.1}$$

The use of a martingale for the description of the price dynamics naturally arises from the hypothesis of efficient market and from the empirical evidence of the absence of simple autocorrelation between price increments. The consequence of this kind of model for the price is that is impossible to extract any information on the future price increments from an analysis of the past increments.

The technical analysis is the study of the market behavior underpinned on the inspection of the price graphs. The technical analysis permits the speculation of the

M. Cristelli, *Complexity in Financial Markets*,
Springer Theses, DOI: 10.1007/978-3-319-00723-6_8,

future value of the price. According to the technical approach the analysis of the past prices can lead to the forecast of the future value of prices. This approach is based upon three basic assumptions [2]:

1. **the market discounts everything**: the price reflects all the possible causes of the price movements (investors' psychology, political contingencies and so on) so the price graph is the only tool to be considered in order to make a prevision.
2. **price moves in trends**: price moves as a part of a trend, which can have three direction: up, down, sideways. According to the technical approach, a trend is more likely to continue than to stop. The ultimate goal of the technical analysis is to spot a trend in its early stage and to exploit it investing in its direction.
3. **history repeats itself**: Thousands of price graphs of the past have been analyzed and some figures (or patterns) of the price graphs have been linked to an upward or downward trend [2]. The technical analysis argues that a price trend reflects the market psychology. The hypothesis of the technical analysis is that if these patterns anticipated a specific trend in the past they would do the same in the future. The psychology of the investors do not change over time therefore an investor would always react in the same way when he undergoes the same conditions.

One reason for the technical analysis to work could be the existence of a feedback effect called *self-fulfilling prophecy*. Financial markets have a unique feature: the study of the market affects the market itself because the results of the studies will be probably used in the decision processes by the investors[1]. The spread of the technical analysis entails that a large number of investors have become familiar with the use of the so called *figures*. A figure is a specific pattern of the price associated to a future bullish or bearish trend. Therefore, it is believed that a large amount of money have been moved in reply to bullish or bearish *figures* causing price changes. In a market, if a large number of investors has the same expectations on the future value of the price and they react in the same way to this expectation they will operate in such a way to fulfill their own expectations. As a consequence, the theories that predicted those expectation will gain investors' trust triggering a positive feedback loop. In this chapter we tried to measure quantitatively the trust on one of the figures of technical analysis.

8.1.2 Supports and Resistances

Let us now describe a particular figure: supports and resistances. The definition of support and resistance of the technical analysis is rather qualitative: a support is a price level, local *minimum* of the price, where the price will bounce on other times afterward while a resistance is a price level, local *maximum* of the price, where the price will bounce on other times afterward. We expect that when a substantial number of investors detect a support or a resistance the probability that the price bounces

[1] Other disciplines such as physics do not have to face this issue.

on the support or resistance level is bigger than the probability the price crosses the support or resistance level. Whether the investors regard a local minimum or maximum as a support or a resistance or not can be related to: (1) the number of previous bounces on a given price level, (2) the time scale. The investors could *a priori* look at heterogeneous time scales. This introduces two parameters which we allow to vary during the analysis in order to understand if and how they affect our results.

8.2 Supports and Resistances: Quantitative Definition

One has to face two issues trying to build a quantitative definition of support and resistance:

1. We define a bounce of the price on a support/resistance level as the event of a future price entering in a stripe centered on the support/resistance and exiting from the stripe without crossing it. Furthermore, we want to develop a quantitative definition compatible with the way the investors use to spot the support/resistances in the price graphs. In fact the assumed memory effect of the price stems from the visual inspection of the graphs that comes before an investment decision. To clarify this point let us consider the three price graphs in Fig. 8.1. The graph in the top panel shows the price tick-by-tick of British Petroleum in the 18th trading day of 2002. If we look to the price at the time scale of the blue circle we can state that there are two bounces on a resistance, neglecting the price fluctuation in minor time scales. Conversely, if we look to the price at the time scale of the red circle we can state that there are three bounces on a resistance, neglecting the price fluctuation in greater time scales. The bare eye distinguishes between bounces at different time scales. Therefore we choose to analyze separately the price bounces at different time scales. To select the time scale to be used for the analysis of the bounces, we considered the time series $P_\tau(t_i)$ obtained picking out a price every τ ticks from the time series *tick-by-tick*[2]. The obtained time series is a subset of the original one: if the latter has N terms then the former has $[N/\tau]$ terms[3]. In this way we can remove the information on the price fluctuations for time scales less than τ. The two graphs in Fig. 8.1 (bottom panel) show the price time series obtained from the *tick-by-tick* recordings respectively every 50 and 10 ticks. We can see that the blue graph on the left shows only the bounces at the greater time scale (the time scale of the blue circle) as the price fluctuations at the minor time scale (the one of the red circle) are absent. Conversely these price fluctuations at the minor time scale are evident in the red graph on the right.
2. The width δ of the stripe centered on the support or resistance at the time scale τ is defined as

[2] We have a record of the price for every operation.

[3] The square brackets [ ] indicate the floor function defined as $[x] = \max\{m \in \mathbf{Z} | m \leq x\}$.

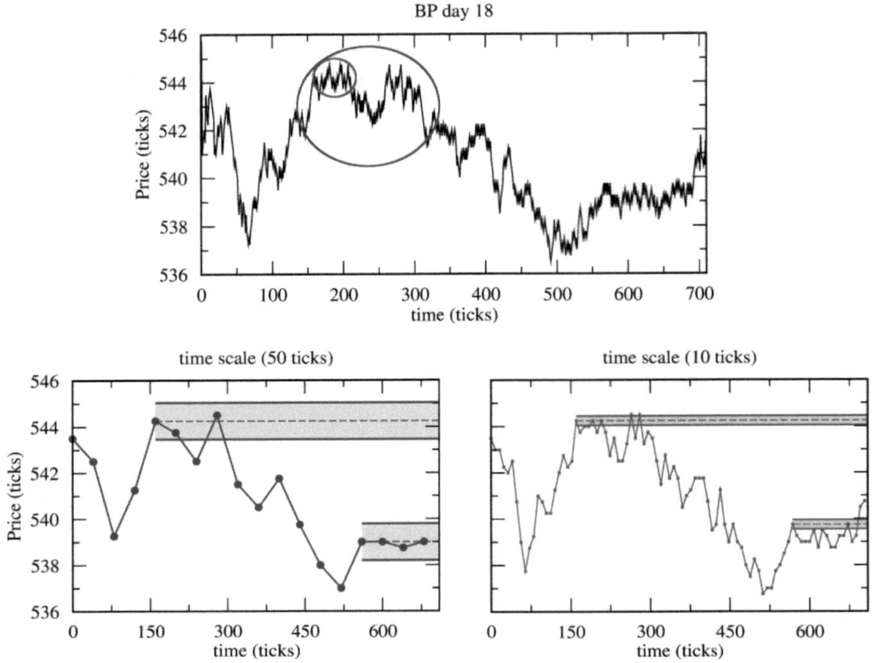

Fig. 8.1 The graph above illustrates the price (in *black*) *tick-by-tick* of the stock British Petroleum in the 18th trading day of 2002. The *blue* and *red circles* define two regions of different size where we want to look for supports and resistances. The graph below in the *left* shows the price (in *blue*) extracted from the time series *tick-by-tick* picking out a price every 50 ticks in the same trading day of the same stock. The graph below in the *right* shows the price (in *red*) extracted from the time series *tick-by-tick* picking out a price every 10 ticks. The *horizontal lines* represent the stripe of the resistance to be analyzed

$$\delta(\tau) = \left[\frac{\tau}{N}\right] \sum_{k=1}^{[\frac{\tau}{N}]-1} |P_\tau(t_{k+1}) - P_\tau(t_k)| \tag{8.2}$$

that is the average of the absolute value of the price increments at time scale τ. Therefore δ depends both on the trading day and on the time scale and it generally rises as τ does. In fact it approximately holds that $\delta_\tau \sim \tau^\alpha$ where α is the diffusion exponent of the price in the day considered. The width of the stripe represents the tolerance of the investors on a given support or resistance: if the price drops below this threshold the investors regard the support or resistance as broken.

To sum up, we try to separate the analysis of the bounces of price on supports and resistances for different time scales. Provided this quantitative definition of support and resistance in term of bounces we perform an analysis of the bounces in order to determine if there is a memory effect on the price dynamics on the previous bounces and if this effect is statistically meaningful.

8.3 Empirical Evidence of Memory Effects

The analysis presented in this chapter are carried out on the high frequency (*tick-by-tick*) time series of the price of 9 stocks of the London Stock Exchange in 2002, that is 251 trading days. The analyzed stocks are: AstraZeNeca (AZN), British Petroleum (BP), GlaxoSmithKline (GSK), Heritage Financial Group (HBOS), Royal Bank of Scotland Group (RBS), Rio Tinto (RIO), Royal Dutch Shell (SHEL), Unilever (ULVR), Vodafone Group (VOD).

The price of these stocks is measured in *ticks*[4]. The time is measured in seconds. We choose to adopt the physical time because we believe that investors perceive this one. We checked that the results are different as we analyze the data with the time in ticks or in seconds. In addition to this a measure of the time in *ticks* would make difficult to compare and aggregate the results for different stocks. In fact, while the number of seconds of trading does not change from stock to stock the number of operation per day can be very different.

We measure the conditional probability of bounce $p(b|b_{prev})$ given b_{prev} previous bounces. This is the probability that the price bounces on a local maximum or minimum given b_{prev} previous bounces. Practically, we record if the price, when is within the stripe of a support or resistance, bounces or crosses it for every day of trading and for every stock. We assume that all the supports or resistances detected in different days of the considered year are statistically equal. As a result we obtain the bounce frequency $f(b|b_{prev}) = \frac{n_{b_{prev}}}{N}$ for the total year. Now we can estimate $p(b|b_{prev})$ with the method of the Bayesian inference: we infer $p(b|b_{prev})$ from the number of bounces $n_{b_{prev}}$ and from the total number of trials N assuming that n is a realization of a Bernoulli process because when the price is contained into the stripe of a previous local minimum or maximum it can only bounce on it or cross it.

Using this framework we can evaluate the expected value and the variance of $p(b|b_{prev})$ using the Bayes theorem [3, 4]:

$$E[p(b|b_{prev})] = \frac{n_{b_{prev}} + 1}{N + 2} \tag{8.3}$$

$$Var[p(b|b_{prev})] = \frac{(n_{b_{prev}} + 1)(N - n_{b_{prev}} + 1)}{(N + 3)(N + 2)^2} \tag{8.4}$$

In Figs. 8.2 and 8.3 the conditional probabilities are shown for different time scales. The data of the stocks have been compared to the time series of the shuffled returns of the price. In this way we can compare the stock data with a time series with the same statistical properties but without any memory effect. As shown in the graphs, the probabilities of bounce of the shuffled time series are nearly 0.5 while the probabilities of bounce of the stock data are well above 0.5. In addition to this, it is noticeable that the probability of bounce rises up as b_{prev} increases. Conversely, the probability of bounce of the shuffled time series is nearly constant. The increase of $p(b|b_{prev})$

[4] A *tick* is the minimum change of the price.

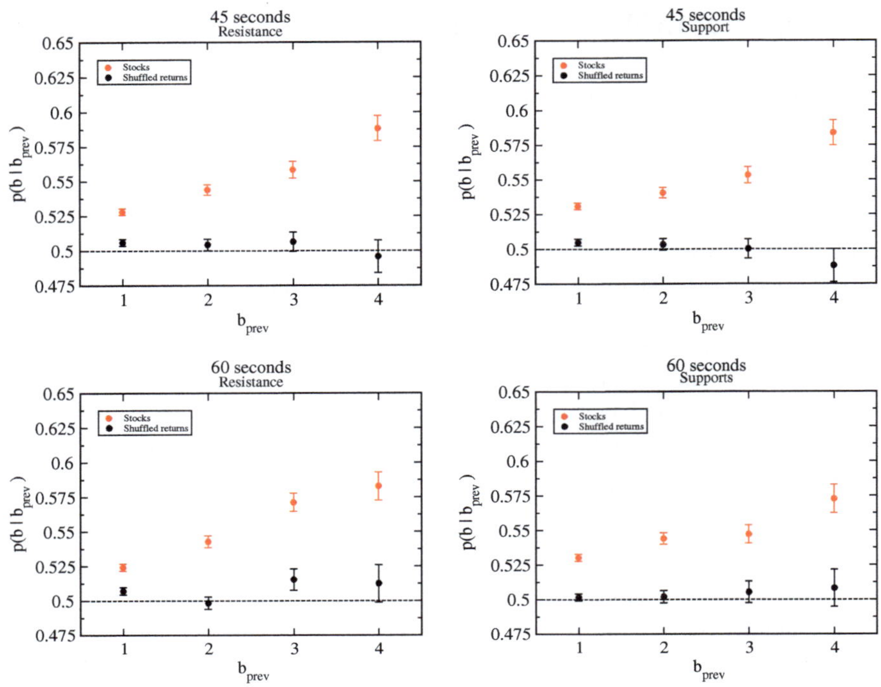

Fig. 8.2 Graphs of the conditional probability of bounce on a resistance/support given the occurrence of b_{prev} previous bounces. Time scale: $\tau = 45, 60\,\mathrm{s}$. The data refers to the 9 stocks considered. The data of the stocks are shown as *red circles* while the data of the time series of the shuffled returns of the price are shown as *black circles*. The graphs in the *left* refer to the resistances while the ones on the *right* refer to the supports

of the stocks with b_{prev} can be interpreted as the growth of the investors' trust on the support or the resistance as the number of bounces grows. The more the number of previous bounces on a certain price level the stronger the trust that the support or the resistance cannot be broken soon. As we outlined above, a feedback effect holds therefore an increase of the investors' trust on a support or a resistance entails a decrease of the probability of crossing that level of price.

We have performed a χ^2 test to verify if the hypothesis of growth of $p(b|b_{prev})$ is statistically meaningful. The independence test ($p(b|b_{prev}) = c$) is performed both on the stock data and on the data of the shuffled time series and we compute

$$\chi^2 = \frac{\sum_{b_{prev}=1}^{4} \left[p(b|b_{prev}) - c\right]^2}{\sum_{b_{prev}=1}^{4} \sigma_{b_{prev}}^2}.$$

Then we compute the p-value associated to a χ^2 distribution with 2 degrees of freedom. We choose a significance level $\alpha = 0.05$. If p-value $< \alpha$ the independence

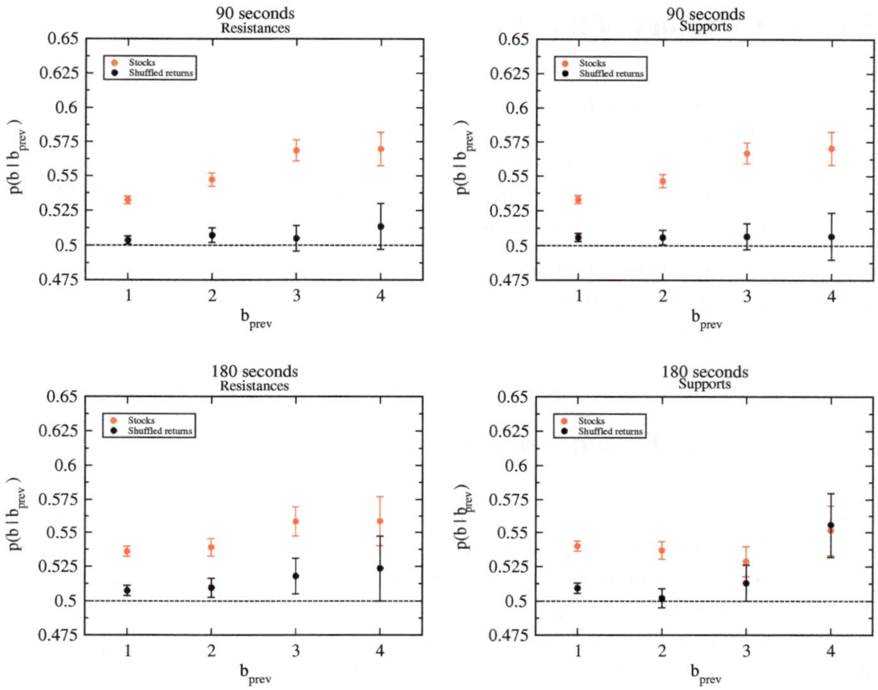

Fig. 8.3 Graphs of the conditional probability of bounce on a resistance/support given the occurrence of k previous bounces. Time scale: $\tau = 90$, 180 s. The data refers to the 9 stocks considered. The data of the stocks are shown as *red circles* while the data of the time series of the shuffled returns of the price are shown as *black circles*. The graphs in the *left* refer to the resistances while the ones on the *right* refer to the supports

Table 8.1 The table shows the p-values for the stock data and for the time series of the shuffled returns for different time scale and for the supports and resistances. The italic cells indicate independence of $p(b|b_{prev})$ to the b_{prev} value. The bold cells indicate a non trivial dependence of $p(b|b_{prev})$ to the b_{prev} value

		45 s	60 s	90 s	180 s
Stocks	Resistances	*<0.0001*	*<0.0001*	*<0.0001*	**0.077**
Stocks	Supports	*<0.0001*	*<0.0001*	*<0.0001*	**0.318**
Shuffled returns	Resistances	**0.280**	**0.051**	**0.229**	**0.583**
Shuffled returns	Supports	**0.192**	**0.229**	**0.818**	**0.085**

hypothesis is rejected while if p-value $\geq \alpha$ it is accepted. The results are shown in Table 8.1. The green cells indicate independence of $p(b|b_{prev})$ on the value of b_{prev} while the red cells indicate dependence of $p(b|b_{prev})$. The results show that there is a clear increase of the investors' memory on the supports/resistances as the number of previous bounces increases for the time scales of 45, 60 and 90 s. Conversely, this memory do not increase at the time scale of 180 s.

8.4 Long Memory of the Price

The analysis of the conditional probability $p(b|b_{prev})$ proves the existence of a long memory in the price time series. We used the Hurst exponent H as a measure of a such long term memory or autocorrelation. The Hurst exponent is estimated via the detrended fluctuation method [5, 6]. It is useful to recall the information that the Hurst exponent provides about the autocorrelation of the time series:

- if $H < 0.5$ one has negative correlation and antipersistent behavior
- if $H = 0.5$ one has no correlation
- if $H > 0.5$ one has positive correlation and persistent behavior.

8.4.1 Empirical Evidence of the Anticorrelation

If we consider the graph in Fig. 8.4a it is noticeable that the price increments are anticorrelated in the given day being the slope of the linear fit $H = 0.44 < 0.5$. The process is not anticorrelated every day: we find $H > 0.5$ in some days and $H < 0.5$ in others. However the average Hurst exponent is $\langle H \rangle < 0.5$ therefore the price increments are anticorrelated on average. The graph 8.4b shows the histogram of H measured in the 251 trading days of 2002 for RIO. The Table 8.2 shows the average values of the Hurst exponent for all the 9 stocks analyzed in this chapter. The table shows that $\langle H \rangle$ is always less than 0.5 therefore there is anticorrelation effect of the price increments for the 9 stocks analyzed.

The anticorrelation of the price increments could lead to an increase of the bounces and therefore it could mimic a memory of the price on a support or resistance. We perform an analysis of the bounces on a antipersistent fractional random walk to verify if the memory effect depends on the antipersistent nature of the price in the time scale of the day. We choose a fractional random walk with the Hurst exponent $H = \langle H_{stock} \rangle$ given by the average over the H exponents of the different stocks shown in Table 8.2. The result is shown in Fig. 8.5. The conditional probabilities $p(b|b_{prev})$ are very close to 0.5 although above this value. In addition to this it is clear that $p(b|b_{prev})$ is constant. These two results prove that the memory effect of the price does not depend on its antipersistent properties, or at least the antipersistence is not the main source of this effect.

8.5 Features of the Bounces

We now want to describe two features of the bounces: the time occurring between two consecutive bounces and the maximum deviation of the price from the support or resistance between two consecutive bounces.

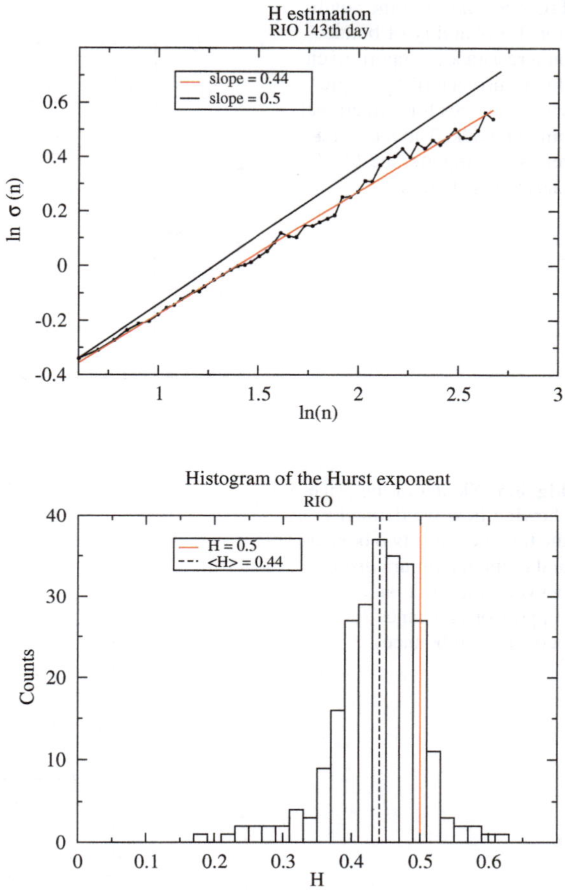

Fig. 8.4 (*top*) Graph of $\ln\sigma(n)$ against n of a trading day of RIO. The *red line* is a linear fit of the data. Its slope gives an estimation of the Hurst exponent. The *black line* represents the linear fit that one would obtain for an uncorrelated time series. (*bottom*) Histogram of H measured in the 251 trading days of 2002 for RIO. The *dotted line* is the average Hurst exponent over the year 2002 while the *red line* indicates $H = 0.5$

Table 8.2 Average values of the Hurst exponent over the year 2002 for all the 9 stocks analyzed in this chapter

Stock	H
AZN	0.471
BP	0.440
GSK	0.464
HBOS	0.472
RBS	0.483
RIO	0.440
SHEL	0.445
ULVR	0.419
VOD	0.419

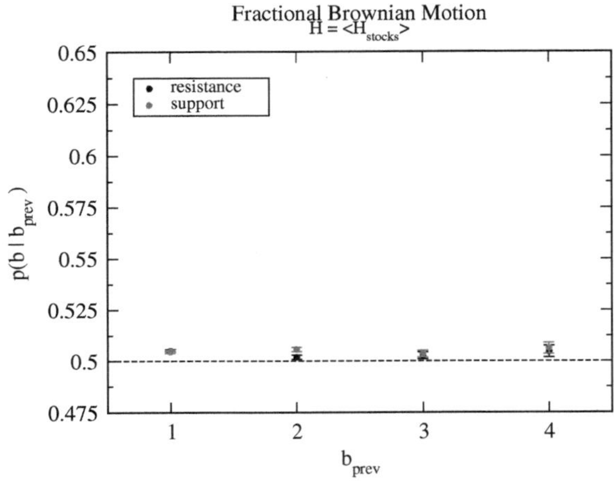

Fig. 8.5 Graph of the conditional probability of bounce on a resistance/support given the occurrence of b_{prev} previous bounces for a fractional random walk. The *red circles* refers to supports, the *black* ones to resistances

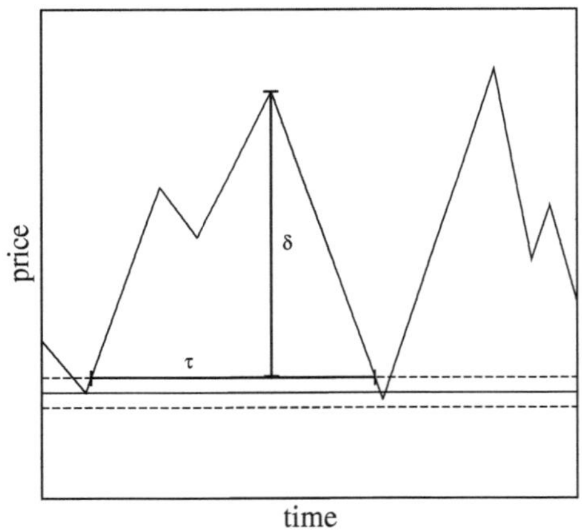

Fig. 8.6 Sketch of the price showing how we defined τ, the time between two bounces and δ, the maximum distance between the price and the support or resistance level between two bounces

We call τ the time occurring between two consecutive bounces. It is defined as the time between an exit of price from the stripe centered on the support or resistance and the subsequent entrance of the price in the same stripe as shown in Fig. 8.6. We make an histogram of τ for different time scales for the stocks and for the shuffled time series as showed in Fig. 8.7. There is no significant difference between the histogram of the stocks and the one of the shuffled time series. τ is measured in terms of the considered time scale so we can compare the histograms at different time scales. It is noticeable that the histograms at different time scale are very similar. We find that a power law fit well describes the histograms of τ.

We call δ the maximum distance in price between two consecutive bounces. The Fig. 8.6 shows how δ is defined. We make an histogram of δ for different time scales for the stocks and for the shuffled time series as showed in Fig. 8.8. In this case a power law fit does not describe accurately the histogram of δ. The histograms related

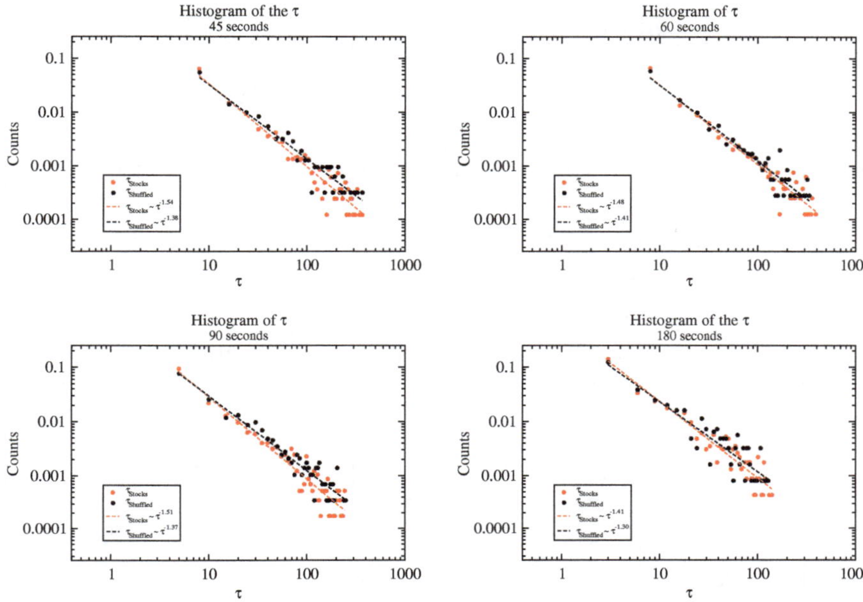

Fig. 8.7 Graphs of the histograms of τ for the time scales of 45, 60, 90, 180 s. We obtained the histograms from the data of all the 9 stocks analyzed in this chapter. We do not make any difference between supports and resistances in this analysis. The *red circles* are related to the stocks while the *black* ones are related to the shuffled time series. The *red dotted line* is a power law fit of the stocks data, the *black dotted line* is a power law fit of the shuffled time series data. The time τ is measured in terms of the considered time scale

to the stocks data and the one related to the shuffled time series are similar in shape. In both cases the distribution of the shuffled case appears to be fatter-tailed than the normal one, this evidence can be interpreted as an additional memory effect of the support/resistance, the price tends to less drift away from the resistance/support value with respect to the shuffled series and consequently the average time between two successive bounces is smaller.

8.6 Summary and Perspectives

Chartist strategies exist and produce detectable signals as shown in this chapter. In the case of resistances and supports we find that the probability of bounces on these values is higher than $1/2$ or anyway is higher than an equivalent random walk or of the shuffled series. In particular we find that the more the number of bounces on these values increases, the more the probability of bouncing on them is higher. This means that the probability of bouncing on a support or a resistance is an increasing

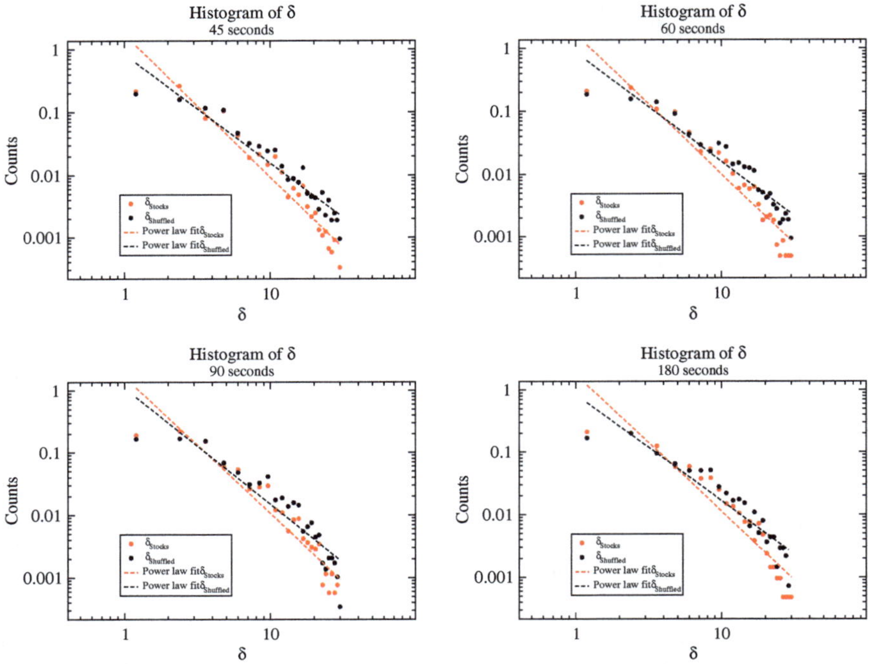

Fig. 8.8 Graphs of δ for the time scales of 45, 60, 90, 180 s. We obtained the histograms from the data of all the 9 stocks analyzed in this chapter. We do not make any difference between supports and resistances in this analysis. The *red circles* are related to the stocks while the *black* ones are related to the shuffled time series. The *red dotted line* is a power law fit of the stocks data, the *black dotted line* is a power law fit of the shuffled time series data. The price δ is measured in ticks

function of the number of previous bounces differently from a random walk or from the shuffled time series in which this probability is independent on the number of previous bounces.

This features is a very interesting quantitative evidence for a self-reinforcement of agents' sentiment, in this case, of the strength of the resistance/support. The more the agents observe bounces the more they expect that the price will again bounce on that value.

It is worth noticing that this finding is in principle an arbitrage opportunity because, once the support or the resistance is detected, the next time the price will be in the nearby of the value a re-bounce will be more likely than the crossing of the resistance/support. In theory it could be possible to create a trading strategy with positive returns and without risk. However, when transaction costs and frictions (i.e. the delay between order submissions and executions) are taken into account, these minor arbitrage opportunities are usually no more profitable. Therefore we intend to verify if the asymmetry in the crossing probability of a support and resistance can give rise to profitable trading system not only in a frictionless market.

References

1. Mantegna, R. N., & Stanley, H. (2000). *An introduction to econophysics: Correlation and complexity in finance*. New York: Cambridge University Press.
2. Murphy, J. (1999). *Technical analysis of the financial markets*. New York: New York institute of finance.
3. Feller, W. (1968). *An introduction to probabbility and its applications* (Vol. 1). New York: Wiley.
4. D'Agostini, G. (2003). *Bayesian reasoning in data analysis: A critical introduction*. Singapore: World Scientific Publishing.
5. Peng, C.-K., Buldyrev, S., Havlin, S., Simons, M., Stanley, H., & Goldberger, A. (1994). Mosaic organization of DNA nucleotides. *Physical Revolution*, *49*, 1685.
6. C.K. Peng, H. S., Havlin, S., & Goldberger, A. (2005). Quantification of scaling exponents and crossover phenomena in nonstationary heartbeat time series. CHAOS, 1.

Chapter 9
Universal Relation Between Skewness and Kurtosis in Complex Dynamics

In this chapter we discuss the work presented in the paper [1] where we identify an important correlation between skewness and kurtosis for a broad class of complex dynamics and present a specific analysis of earthquake and financial time series. We highlight that two regimes of non Gaussianity can be identified: a parabolic one, which is common in various fields of physics, and a power law one, with exponent $4/3$, which, at the moment, appears to be specific of earthquakes and financial markets. For this new property (i.e. the $4/3$ regime) we propose a model and an interpretation in terms of very rare events dominating the statistics independently on the nature of the events considered. The predicted scaling relation between skewness and kurtosis perfectly matches the experimental pattern of the second regime. Regarding the price fluctuations, this situation characterizes a new and universal Stylized Fact.

9.1 Introduction: Skewness and Kurtosis

Deviations from Gaussianity in time series are usually investigated by introducing moments of order higher than the second, which are instead trivial for a normal distribution. It is also a well-known empirical evidence of complex time series that fluctuations are not Gaussian distributed. Here we focus our attention on the statistics of earthquake magnitudes [2] and price variations, often called returns, [3–5] but we stress that our conclusions are *independent* on the nature of the analyzed time series.

Let us briefly discuss the mathematical meaning of skewness and kurtosis. The third normalized moment S, also called skewness, quantifies the asymmetry of the probability density function (pdf). In formula

$$S = E\left[\left(\frac{X-\mu}{\sigma}\right)^3\right] = \frac{\mu_3}{\sigma^3} \tag{9.1}$$

M. Cristelli, *Complexity in Financial Markets*,
Springer Theses, DOI: 10.1007/978-3-319-00723-6_9,

where X is the random variable under consideration, μ is the expected value of this random variable, σ^2 is its variance and μ_3 is its third centered moment. For instance for the earthquake catalog analyzed in this chapter we find that the skewness is always positive because of the lack of negative events, while for price returns the skewness is typically negative ($S < S_{Gauss} = 0$), evidencing that more negative events than positive ones are observed, especially in the tails.

On the other hand the fourth normalized moment K can be seen as a measure of the deviation of the distribution of the random variable from a Gaussian and in formula kurtosis reads as

$$K = E\left[\left(\frac{X - \mu}{\sigma}\right)^4\right] = \frac{\mu_4}{\sigma^3} \tag{9.2}$$

where μ_4 is the fourth centered moment of the random variable X. In both cases here studied the kurtosis is larger than 3 ($K > K_{Gauss} = 3$), pointing out that there are more observations at the mean and in the tails than a normal distribution.

9.2 Empirical Results: Parabolic and 4/3 Regime

One is naturally led to wonder whether a relation between the empirical values of S and K exists. This question is far from being trivial, as different moments refer to different properties of the probability density function. Experimental data from various fields of physics support this hypothesis: a parabolic relation between skewness and kurtosis holds in plasma physics [6], meteorology [7] and oceanography [8]. In these cases, in order to have different points in the $S - K$ plane, the authors repeat the experiment or evaluate the moments on different time windows of the same series.

In [1] we identify a new scaling relation between skewness and kurtosis in earthquake and financial data. Moreover, this relation turns to be universal, in the sense that all the analyzed time series follow the same functional behavior in the $S - K$ plane (it is worth noticing that this statistical feature could be, in principle, found also in other fields of Physics). Regarding price fluctuations this observation is in contrast with a known (albeit discussed) empirical evidence of non universality in the other statistical properties of financial time series [3, 5].

In particular, our financial data clearly exhibit, in contrast with other physical data, two regimes: parabolic near the Gaussian region $S \approx 0$, $K \approx 3$ and power law with exponent 4/3 for large values of the moments. The earthquake data, in contrast, smoothly approach the second regime. In addition the ubiquity of this evidence calls for a general explanation which we are going to discuss in the next section in which we propose a simple argument for the second regime, having noticed that the values of the moments are mainly due to the largest event of the sample. Using this concept, we are able to derive a power law relation between skewness and kurtosis which shows an excellent agreement with the empirical data.

We used three datasets in our analysis. The first earthquake dataset is taken from the ANSS catalog (www.ncedc.org/anss) and covers the magnitudes of all the events registered from 01/01/1990 to 17/05/2011. The second dataset is taken from the ISIDe catalog (iside.rm.ingv.it) and includes the Italian earthquakes from 16/04/2005 to 04/05/2011. For our statistical analysis we consider the energy of the ith earthquake $r_i = 10^{1.5m_i}$, where m_i is the registered magnitude. The datasets are divided in subsamples of length N. The ANSS catalog includes 10741 subsamples and the ISIDe catalog 515 subsamples, each made of $N = 100$ events. Our financial dataset consists in the daily price returns $r_i = \log p_i - \log p_{i-1}$ where p_i is the closing price of the day i. We consider eight NYSE stocks traded for 25 up to 48 years, plus the S&P 500 index (60 years of trading). The data is taken from finance.yahoo.com. For financial data we choose $N = 250$, so that each subsample covers about one year. We have about 400 subsamples.

We calculate the normalized third and fourth moment (skewness and kurtosis) from their sample averages as $S = \left(\frac{1}{N}\sum_{i=1}^{N}(r_i - \mu)^3\right)/\sigma^3$ and $K = \left(\frac{1}{N}\sum_{i=1}^{N}(r_i - \mu)^4\right)/\sigma^4$, where μ is the sample mean and σ is the standard deviation, for each subsample.

In Figs. 9.1 and 9.2 we plot the kurtosis as a function of the skewness for earthquake and financial data, respectively. We use different symbols for the two catalogs, for each stock and the index to stress the fact that, despite the distinct origins of the samples, we always find the same relation between skewness and kurtosis, as points from different samples all follow the same pattern. We point out that no rescaling procedure has been performed.

In Fig. 9.1 the points cluster in a parabolic region. To understand this behavior we first focus on financial data.

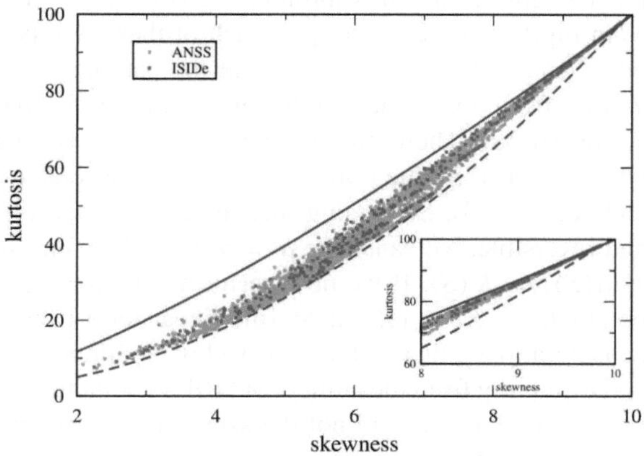

Fig. 9.1 Kurtosis versus skewness for the energies of the earthquakes in the ANSS and ISIDe catalogs. The *broken line* is the lower bound in [9], while the *solid one* is the prediction of Eq. 9.5, which matches the empirical points for large values of S and K. *Inset* zoom of the high moment region

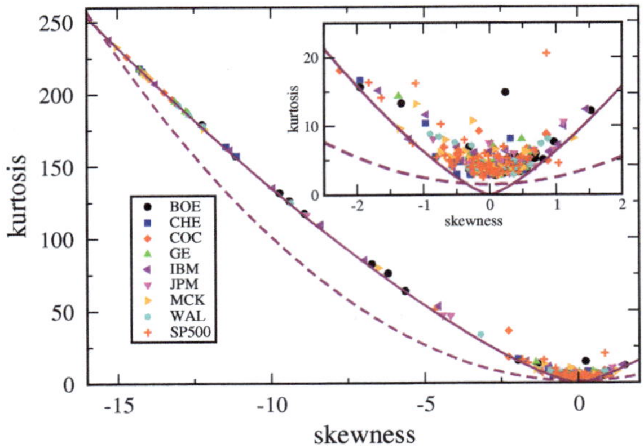

Fig. 9.2 Kurtosis versus skewness for the daily returns of a series of stocks and one index. Each point refers to a sample of about one year. The *broken line* is the lower bound in [9]. The *solid line* is the prediction of Eq. 9.5, which matches very well the empirical points corresponding to the samples with large values of S and K. *Inset* zoom of the parabolic region

9.2.1 Parabolic Regime

In Fig. 9.2 we can detect two distinct regimes for the dependence of the kurtosis on the skewness: the dense parabolic cluster around the point $S = 0, K = 3$ which would characterize an infinite Gaussian sample and the left branch characterized by high values of kurtosis and negative skewness. A recent work by Sattin et al. [10] gives a simple explanation of the parabolic dispersion of the first regime. The authors observe that this parabolic dependence is a common feature among time series deriving from several fields of physics (see [10] and references therein). This relation is well-known but its origin is still a matter of debate. Their argument goes as follows. Let us consider a time series which we divide in subsamples of a certain fixed length. Each of this subsample gives rise to a point in a skewness-kurtosis plane as in Fig. 9.1. We can interpret each of these points as generated by an underlying pdf $f(x|a)$ where we suppose that the parameter a characterizes the deviation of $f(x|a)$ from a Gaussian. In formula, $\lim_{a\to 0} f(x|a) = \mathcal{N}(0,1)$ where $\mathcal{N}(0,1)$ is a Normal pdf. Therefore skewness and kurtosis must depend on the parameter a which determines the point $(S(a), K(a))$ associated to each subsample ($S(0) = 0$ and $K(0) = 3$). Supposing that the dependence of the skewness on a is smooth and reversible, we can write $a = a(S)$ and we can substitute it in $K(a)$, obtaining $K(a(S)) = K(S)$. If we now perform a Taylor expansion of K around $S = 0$ we obtain $K = K_0 + (K''/2)S^2$ (the linear term in most cases is zero for symmetry considerations), that is, the observed parabolic relation.

Differently from the authors of [10] we want to briefly discuss why we can Taylor expand the kurtosis and not the skewness. In [10] it is generally argued that if the symmetry condition $f(x) = f(-x)$ holds then the argument above mentioned is valid. However, the question is subtler and, as we are going to show, the correct symmetry the system must satisfied is instead $f(x|a) = f(-x|-a)$. In order to investigate which are the implications of this condition on $S(a)$ and $K(a)$, we introduce the moment generating function $M(q|a)$ [11, 12] of $f(x|a)$, that is the Laplace transform of the pdf, in formula $M(q|a) = \int \exp(qx)\, f(x|a)\, \mathrm{d}x$. Now the symmetry condition reads as $M(q|a) = M(-q|-a)$ in the transformed space. Now recalling that $M(q|a) = \sum_{k=0}^{\infty} m_k(a) q^k/k!$, where $m_k(a) = E(x^k|a)$, and substituting this last expression in the symmetry condition we find the following parity condition for the moments: $m_{2k}(a) = m_{2k}(-a)$ and $m_{2k+1}(a) = -m_{2k+1}(-a)$, so $S(a)$ is an odd function and that $K(a)$ an even one. This implies that around $a = 0$ only $S(a)$ is invertible. Furthermore the well-defined parity of $S(a)$ and $K(a)$ clarifies why the linear term in the final expansion $K = K_0 + (K''/2)S^2$ is missing.

Another possible explanation for the parabolic behavior is linked to the existence of lower bounds for the difference $K - S^2$. The specific value of the bound depends on the shape of the pdf. In 1916 Pearson [13] found that this bound is in general 2, while more recently Klaassen et al. [9] showed that, for unimodal distributions, $K - S^2 \geq \frac{189}{125}$. In our case, the specific shape of the pdf is sample dependent; however, we may suppose that a parabolic lower bound exists. We have verified this assumption by means of extensive Monte Carlo simulations even on (finite) Gaussian time series. This bound is the broken line in Figs. 9.1 and 9.2.

9.2.2 The 4/3 Regime

The mentioned arguments work well for small deviations from Gaussianity while this is not the case for our data. In fact the second regime observed in Figs. 9.1 and 9.2 appears to be very peculiar of financial markets and large earthquakes even if, *a priori*, this behavior could be observed in physical phenomena, as we are going to show. Away from the Gaussian region the relation between skewness and kurtosis is no longer parabolic but becomes a power law with exponent 4/3. We propose the following simple argument in order to explain this new behavior.

Let us suppose that in the time series $\{r_i\}$ there is an extreme event $\overline{r}$. If $\overline{r}$ is sufficiently larger than the sum of the other events in the subsamples we can approximate skewness and kurtosis with the maximum values, that is,

$$S \simeq \frac{\frac{1}{N}(\overline{r} - \mu)^3}{\sigma^3} \tag{9.3}$$

$$K \simeq \frac{\frac{1}{N}(\overline{r} - \mu)^4}{\sigma^4}. \tag{9.4}$$

From Eq. 9.3 we can easily find that $(NS)^{1/3} \simeq (\overline{r} - \mu)/\sigma$; this expression can be substituted in Eq. 9.4 and gives

$$K \simeq N^{1/3} S^{4/3} \tag{9.5}$$

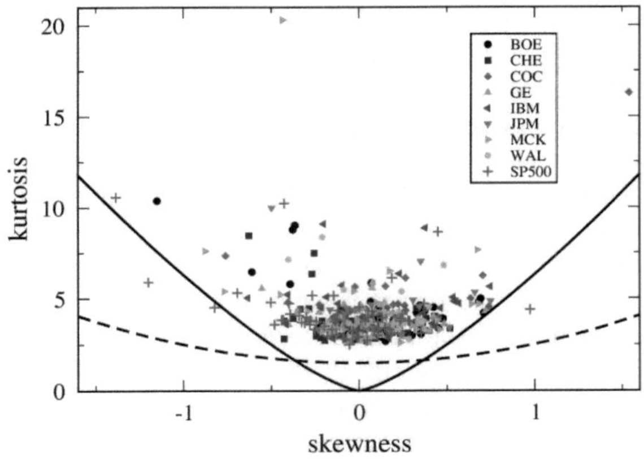

Fig. 9.3 Same plot of Fig. 9.2, but with the largest event removed. Almost all the points shrink in the parabolic region. The *solid line* does no more match the empirical points, while the bound is still valid

where, in our case, we have considered samples of $N = 100$ earthquakes and $N = 250$ trading days. As one can see from Fig. 9.2 (solid line), the agreement of Eq. 9.5 with the financial data is excellent, while for earthquakes the agreement is good only for high values of K. We are going to discuss this discrepancy later in the chapter. Our argument is also supported by Fig. 9.3 where we repeat the same analysis of Fig. 9.2 removing the largest event from each subsample. Now almost all the points lie in the parabolic regime. In order to understand the range of validity of the approximations in Eqs. 9.3 and 9.4 used to derive the scaling relation of Eq. 9.5 we analyze the

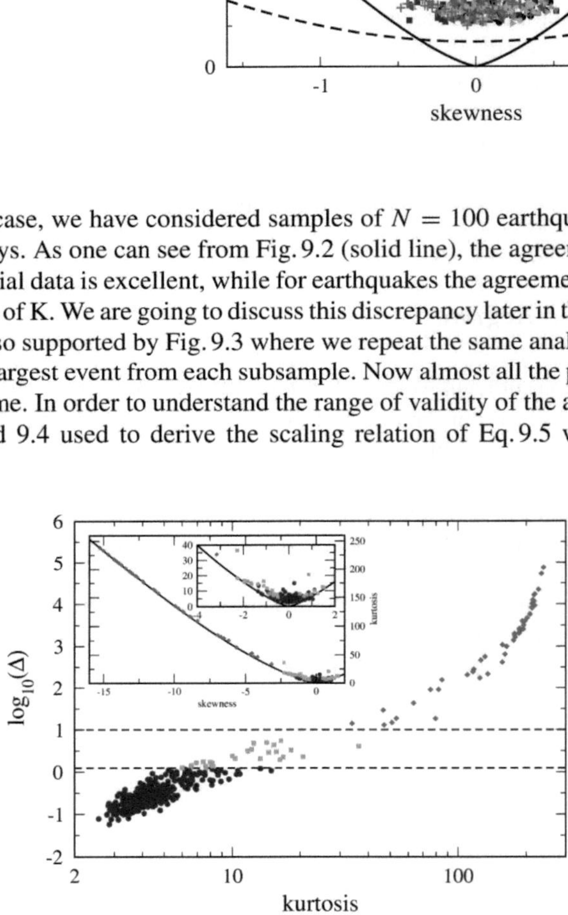

Fig. 9.4 Financial data. The three regions of points characterized by different values of Δ ($\Delta < 1$ *blue circles*, $1 < \Delta < 10$ *green squares*, $\Delta > 10$ *red diamonds*) are mapped in the $S - K$ plane in an univocal way. *Inset* evidence for correlation between the dominance of the maximum event and the kurtosis in the case $N = 250$. Hence the $4/3$ power law regime is due to all and only those points generated from the subsamples which have an event much larger than the sum of the others. In the Gaussian case all points lie in a small region around $K = 3$ (not shown) (Color figure online)

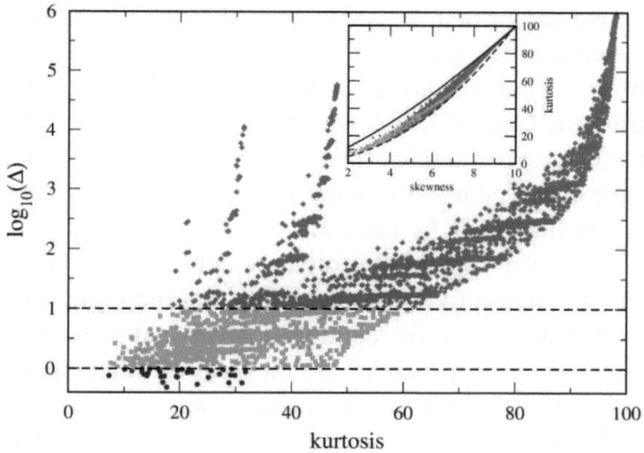

Fig. 9.5 The $\Delta - K$ plot for the earthquake databases. As in Fig. 9.4, there is an univocal mapping from the regions of high Δ and the power law regime in the $S - K$ plot

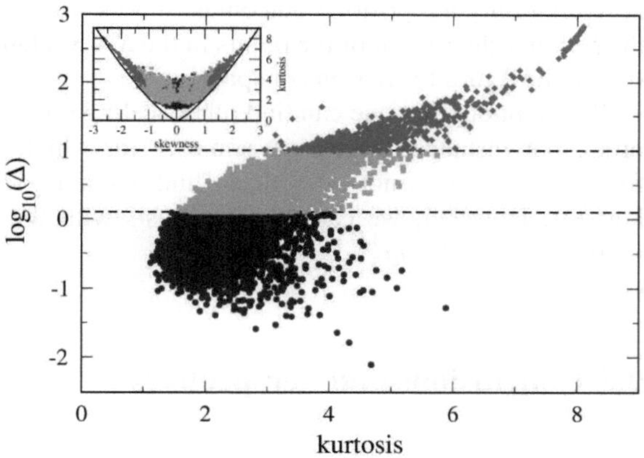

Fig. 9.6 The same plot of Fig. 9.4, but here each subsamples has $N = 10$ elements. Having considered a lower value of N, we do not expect our approximation to be valid. In fact, most of the points have a small kurtosis and lie in the parabolic region. In the Gaussian case we have found a similar behavior (not shown)

limit case in which N events are equal to a constant value δ except one which is equal to $\alpha\delta$ with $\alpha > 1$. In this case the (not normalized) nth moment is equal to $M_n = \frac{1}{N+1}\sum_{i=1}^{N+1} r_i^n = \frac{1}{N+1}(N\delta^n + \alpha^n\delta^n)$ where we have set $\mu = 0$ for the sake of simplicity. In this framework it is easy to see that the approximation is valid if $\alpha^n \gg N$. Let us verify that all and only the samples for which this approximation is valid are characterized by Eq. 9.5. For each subsample we define $\Delta = \bar{r}^4 / \sum_{i=1}^{N-1} r_i^4$ and we plot this quantity versus the kurtosis (see Figs. 9.4 and 9.5). We identify three regions: the Gaussian-like samples, having $\Delta < 1$ (blue circles), an intermediate regime, with $1 < \Delta < 10$ (green squares) and the samples dominated by the largest event, having $\Delta > 10$ (red diamonds). The three regions are univocally mapped into the $S - K$ plane, evidencing that the 4/3 scaling regime is produced only by those samples which have a very large event. On the contrary, all the other points lie in the parabolic region. Moreover we can repeat the very same analysis for smaller subsamples ($N = 10$), for which we do not expect our approximation to be valid. As shown in Fig. 9.6, almost all the points belong to the first regime.

We performed the same analysis for finite Gaussian samples. We do not show the results for reason of space, however we have observed a similar behavior for $N = 10$ but increasing N all points collapse towards $S = 0$, $K = 3$ as expected from the Central Limit Theorem. This clearly indicates that the $S - K$ behavior of the Gaussian samples for $N = 10$ is a finite size effect, differently from what we have observed in earthquake and financial data, where increasing N the points adjust on the line given by Eq. 9.5.

The origin of the discrepancy between earthquake and financial data (Figs. 9.1 and 9.2, respectively) is in the fact that the support of the magnitude pdf is bounded from below, as there cannot be earthquakes with negative energies. This implies an intrinsic distortion in the moment estimate, as the central part of the pdf plays the role of the left tail. As a consequence, given the same kurtosis the skewness of such an asymmetric distribution will be larger with respect to a symmetric one. We have checked this hypothesis by generating samples from a Normal distribution and accepting only the positive outcomes. In this way we have been able to reproduce the peculiar dispersion of the points in the $S - K$ plane shown in Fig. 9.1. We do not shown the figure for reasons of space.

We point out that one can find other scaling relations between moment of higher order. For instance, there exist some N for which the approximation holds only for $n \geq 4$, so it is not possible to find a scaling relation between skewness and kurtosis. However, we can relate kurtosis with the fifth normalized moment m_5 finding $K = N^{-1/5} m_5^{4/5}$.

9.3 Conclusions and Perspectives

In [1] we identify a new regime of the scaling between skewness and kurtosis for those samples whose statistics is dominated by their largest event independently on the nature of these events. In fact we have observed this scaling regime both in physical and social phenomena, respectively in earthquake and financial data. We propose a theory which predicts a scaling exponent equal to 4/3 and a pre-factor $N^{1/3}$, where N is the size of the sample, and which perfectly matches the empirical patterns in the $S - K$ plane.

We also want to briefly discuss an interesting conclusion which can be drawn from the asymmetry between the right branch in Fig. 9.2. It is reasonable to think that markets are completely symmetric with respect to positive or negative price movements, that is, returns are *a priori* symmetric. On the other hand we observe points generated by samples where the largest event is typically negative ($\overline{r} < 0$ so that $S < 0$). Since we have no valid reasons to believe that market mechanisms or market rules change when price movements become very large, we can conclude that the right branch of the skewness-kurtosis plot is not observed only because of the asymmetric agents' perception of price movements which generates the asymmetric pattern in Fig. 9.2. Therefore the left branch is originated only by an exogenous psychological component and not by a structural one. This is not only a new Stylized Fact but the first empirical evidence for a strongly universal behavior in the statistical properties of financial time series. From this point of view, we believe that the introduction of this new evidence will help the desirable empirical validation of agent based models [14, 15].

In the end it is worth noticing that a suitable analysis in the $S - K$ plane can give a possible quantitative method to determine the nature of the largest events, i.e. they are really off statistics (outliers) or they are only rare events which are still compatible with the statistics given by the dataset (black swan). In fact if the largest event is a black swan and the size of the sample is increased, we expect that the point associated to this enlarged set in the $S - K$ plane will tend towards the parabolic region. Instead if the event is a real outlier we expect that the position of the point in the $S - K$ point is substantially independent on the size N of the set to which the outlier belongs. Therefore the analysis of the dynamics of the points in the $S - K$ plane with respect to the size of the set could in principle give a quantitative criterion to distinguish real outliers from black swans.

References

1. Cristelli, M., Zaccaria, A., & Pietronero, L. (2012). Universal relation between skewness and kurtosis in complex dynamics. *Physical Review E*, *85*, 066108.
2. Gutenberg, B., & Richter, C. (1956). *Annales Geophysicae*, *9*, 1.
3. Bouchaud, J.-P., & Potters, M. (2003). *Theory of financial risk and derivative pricing: From statistical physics to risk management*. Cambridge: Cambridge University Press.
4. Mantegna, R. N., & Stanley, H. (2000). *An introduction to econophysics: Correlation and complexity in Finance*. New York, NY: Cambridge University Press.
5. Cont, R. (2001). Empirical properties of asset returns: Stylized facts and statistical issues. *Quantitative Finance*, *1*, 223.
6. Labit, B., Furno, I., Fasoli, A., Diallo, A., Müller, S. H., Plyushchev, G., et al. (2007). Universal statistical properties of drift-interchange turbulence in TORPEX plasmas. *Physical Review Letters*, *98*, 255002.
7. Alberghi, S., Maurizi, A., & Tampieri, F. (2002). Relationship between the vertical velocity skewness and kurtosis observed during sea-breeze convection. *Journal of Applied Meteorology*, *41*, 885.
8. Sura, P., & Sardeshmukh, P. D. (2008). A global view of non-Gaussian SST variability. *Journal of Physical Oceanography*, *38*, 639.
9. Klaassen, C. A. J., Mokveld, P. J., & van Es, B. (2000). Squared skewness minus kurtosis bounded by 186/125 for unimodal distributions. *Statistics and Probability Letters*, *50*, 131.
10. Sattin, F., Agostini, M., Cavazzana, R., Serianni, G., Scarin, P., & Vianello, N. (2009). About the parabolic relation existing between the skewness and the kurtosis in time series of experimental data. *Physica Scripta*, *79*, 045006.
11. Gardiner, C. (1990). *Handbook of stochastic methods: For physics, chemistry and the natural sciences*. Berlin: Springer.
12. Grimmett, G., & Stirzaker, D. (2004). *Probability and random processes*. Oxford: Oxford University Press.
13. Pearson, K. (1916). Mathematical contributions to the theory of evolution. XIX. Second supplement to a memoir on skew variation. *Philosophical Transactions of the Royal Society of London. Series A*, *216*, 429.

14. Cristelli, M., Zaccaria, A., & Pietronero, L. (2011). Critical overview of agent-based models for economics. In *Proceedings of the School of Physics E. Fermi, course CLXXVI, 2010, Varenna*. http://arxiv.org/abs/arXiv:1101.1847.
15. Chakraborti, A., Toke, I. M., Patriarca, M., & Abergel, F. (2011). Econophysics: Empirical facts and agent-based models. Quantitative Finance (in press). http://arxiv.org/abs/arXiv:0909.1974v2.

Chapter 10
Web Queries Can Predict Stock Market Volumes

In this chapter we present how non financial data can be used to track financial activity (see [1] and Fig. 10.1). In details we investigate query log volumes, i.e. the volumes of searches for a specific query done by users in a search engine as a proxy for trading volume and we find that users' activity on Yahoo! search engine anticipate trading volume by one-two days.

This investigation is motivated by the fact that nowadays we live in a computerized and networked society where many of our actions leave a digital trace and affect other people's actions. This has lead to the emergence of a new data-driven research field [2]: mathematical methods of computer science, statistical physics and sociometry provide insights on a wide range of disciplines [3] ranging from social science [4] to human mobility [5]. A recent important discovery is that query volumes (i.e., the number of requests submitted by users to search engines on the www) can be used to track and, in some cases, to anticipate the dynamics of social phenomena. Successful examples include unemployment levels [6], car and home sales [7], and epidemics spreading [8]. For instance, the spreading of diseases such as influenza can be monitored and even anticipated with some days of advance based on the volume of queries related to the word "flu".

Few recent works applied this approach to stock prices [9] and market sentiment [10]. However, it remains unclear if trends in financial markets can be anticipated by the collective wisdom of on-line users on the web. As anticipated in this chapter we show that trading volumes of stocks traded in NASDAQ-100 are correlated with the volumes of queries related to the same stocks. In particular, query volumes anticipate in many cases peaks of trading by one day or more. Our analysis is carried out on a unique dataset of queries, submitted to an important web search engine (Yahoo! search engine), which enable us to investigate also the user behavior. In addition the analysis is twofold. On the one hand, we assess the relation over time between the number of queries ("query volume", hereafter) related to a particular stock and the amount of exchanges over the same stock ("trading volume" hereafter). We do so by means not only of a time-lagged cross-correlation analysis, but also by means of the Granger-causality test. On the other hand, our unique data set allows us to analyze the

M. Cristelli, *Complexity in Financial Markets*,
Springer Theses, DOI: 10.1007/978-3-319-00723-6_10,

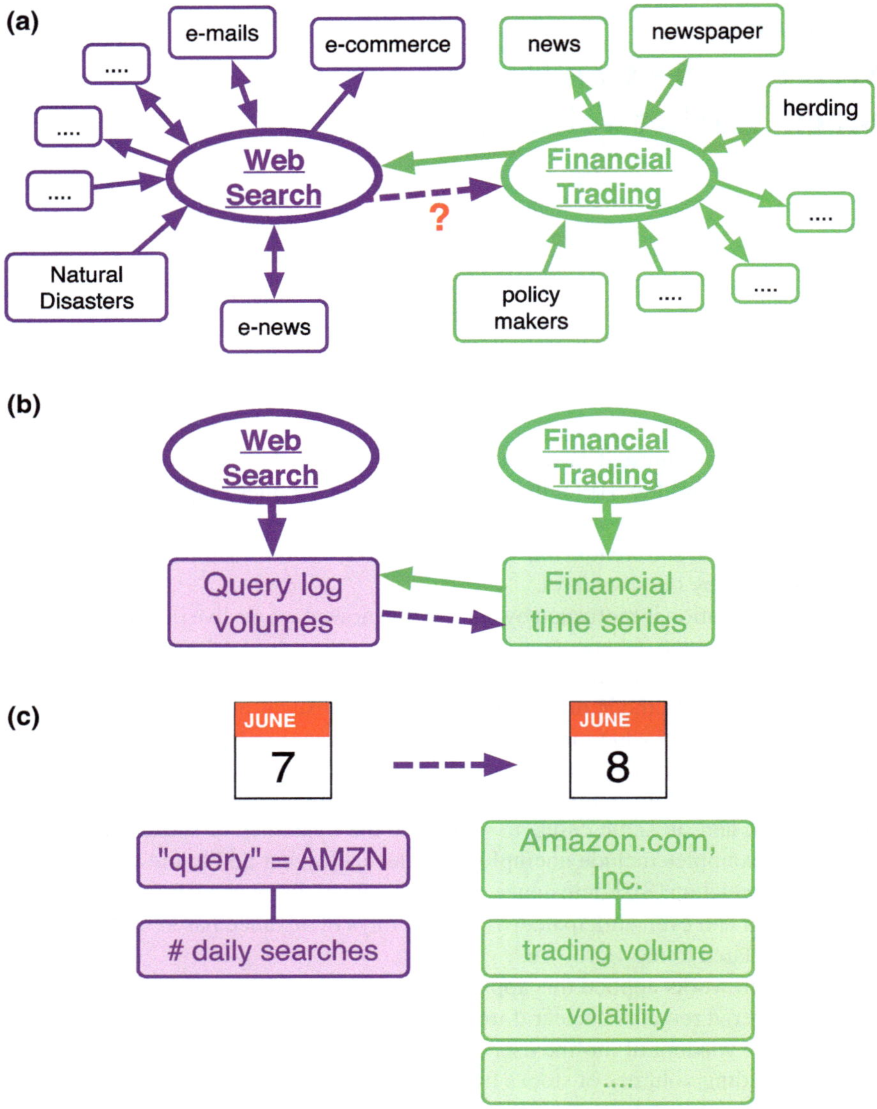

Fig. 10.1 Non financial data can be used to track financial activity

querying activity of individual users in order to provide insights into the emergence of their collective behavior. We want to stress that these findings contribute to the debate on the identification of early warnings of financial systemic risk, based on the activity of users of the www.

10.1 Empirical Evidences

Recent investigations showed that Web search volumes can be used to accurately track several social phenomena [6, 7]. We address the issue whether a similar approach can be applied to obtain early indications of movements in the financial markets. Indeed, financial turnovers, financial contagion and, ultimately, crises, are often originated by collective phenomena such as herding among investors (or, in extreme cases, panic) which signal the intrinsic complexity of the financial system. This is a crucial information since financial turnovers (when crises themselves), can be originated by collective phenomena as herding or in extreme case panic. Collective phenomena are at the basis of the so-called "*domino effects*" present during financial contagion and they signal the intrinsic complexity of the system. Therefore, the possibility to anticipate anomalous collective behavior of investors is of great interest to policy makers because it may allow for a more prompt intervention, when this is appropriate. However, so far, there was no clear evidence that this anticipation is possible.

In our analysis we consider a set of companies ("NASDAQ-100 set" hereafter) that consists of the companies included in the NASDAQ-100 stock market index (the 100 largest non-financial companies traded on NASDAQ). Previous preliminary studies [9] looked at stock prices at a weekly time resolution and found no significant correlation between query volumes and stock prices. In contrast, we look at trading volume at a daily frequency and we find a strong correlation between query volumes and trading volumes for all stocks in the NASDAQ-100 set.

Financial data of daily trading volumes and closure prices are public and have been downloaded from Yahoo! Finance web site. The query-log data analyzed here is a portion of the Yahoo! US search-engine log. It covers a time interval of one year, from May 1st, 2010, to April 30th, 2011. We computed aggregated search volumes for the selected stocks. Each query submitted by a user is stored in the query log together with a timestamp representing the exact time point of its submission. This temporal information can be used to aggregate search volumes at different levels of granularity.

For each traded company, we computed query volumes by extracting two different types of queries: (i) all the queries whose text contained the ticker string as a distinct word (ii) all the queries whose text was exactly matching the company names, after removing the legal ending ("Incorporated" or "Corporation" or "Limited", and all their possible abbreviations). We observed a correlation with the trading volumes when the queries of the first type where used, while no significant results were obtained when the queries matching the company name were taken into consideration. This is probably due to the fact that investors use directly tickers rather than the name of the company when they search for related information in the Web.

In Fig. 10.2 (top panel) we show the time evolution of the query volume of the ticker "NVDA" and the trading volume of the corresponding company stock "NVIDIA Corporation" and Fig. 10.3 (top panel) shows the same plot for query volume of the ticker "RIMM" and the trading volume of company stock "Research In Motion Limited". A simple visual inspection of these figures reveals a clear

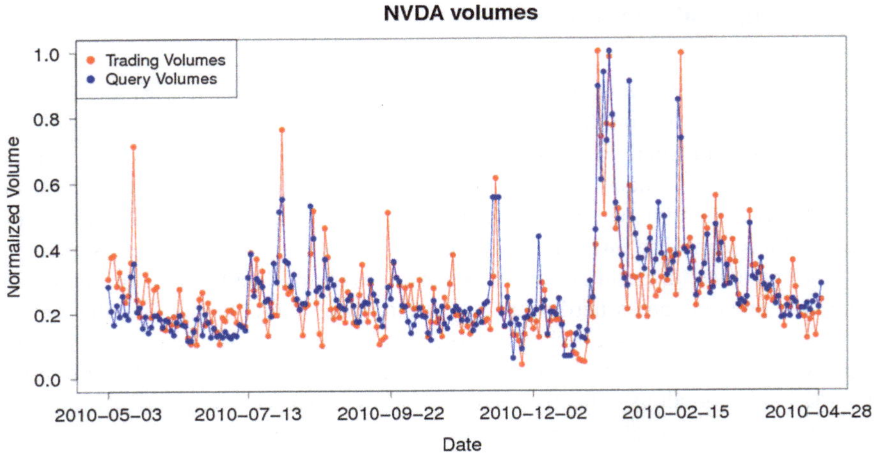

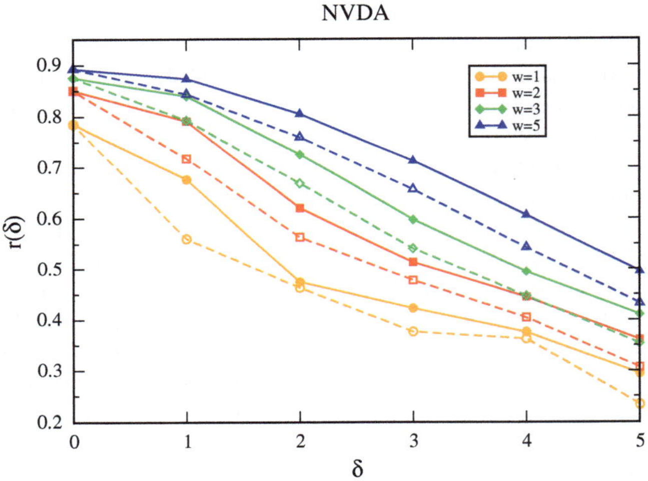

Fig. 10.2 Query log volumes and trading volumes: cross correlation analysis ("NVDA"). (*up*) Time evolution of query-logs normalized volumes for the ticker "NVDA" compared with the trading-volume of the "NVIDIA Corporation". The data for both query-logs (*blue*) and trading volume (*red*) are aggregated on a daily basis. (*bottom*) The plot of the sample cross correlation function $r(\delta)$ as defined in Eq. 10.1 for various values of the weight w of the moving average versus absolute values of the time lag δ (positive values of δ correspond to *solid lines* while negative values of the time lag correspond to the *broken lines*). The correlation coefficients at positive time lags are always larger than the corresponding at negative ones, this suggests that today's web search volumes anticipate and affect the trading activity of the following days (typically one or two days at most)

correlation between the two time series because peaks in one time series tend to occur close to peaks in the other. One could argue that a peak in trading volume tend to trigger a peak of query volume, for instance because anomalous trade volumes

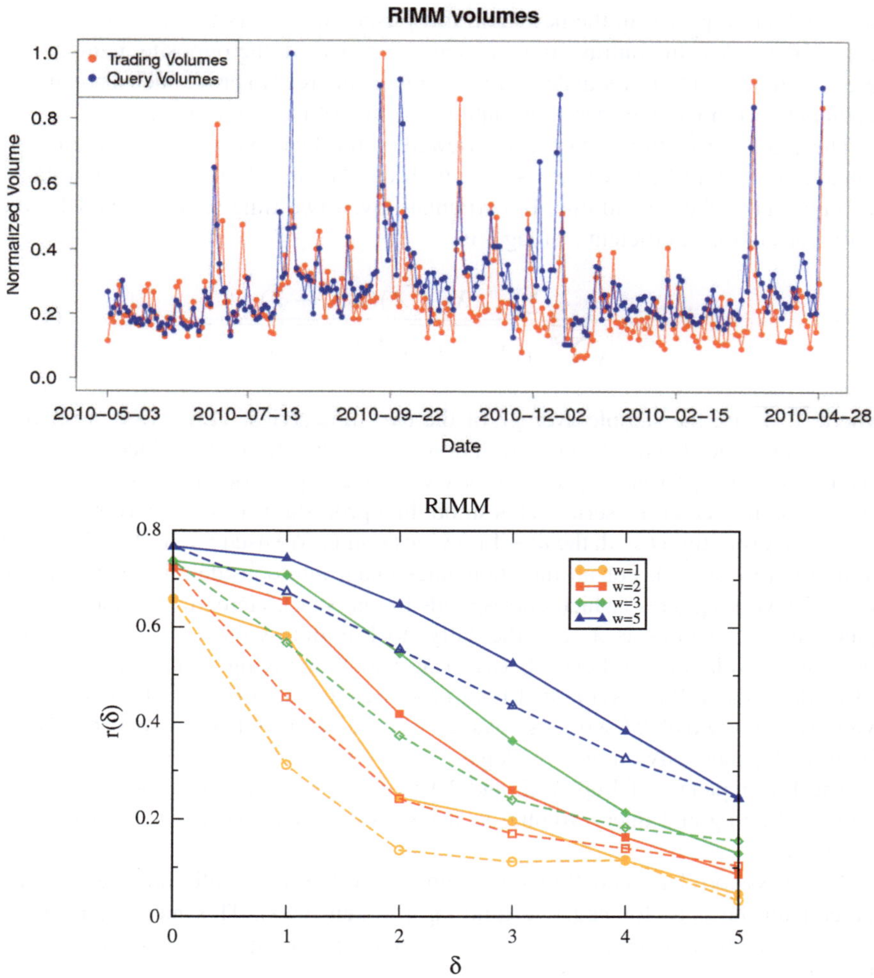

Fig. 10.3 Query log volumes and trading volumes: cross correlation analysis ("RIMM"). (*up*) Time evolution of query-logs normalized volumes for the ticker "RIMM" compared with the trading-volume of the "Research In Motion Limited". The data for both query-logs (*blue*) and trading volume (*red*) are aggregated on a daily basis. (*bottom*) The plot of the sample cross correlation function $r(\delta)$ as defined in Eq. 10.1 for various values of the weight w of the moving average vs absolute values of the time lag δ (positive values of δ correspond to *solid lines* while negative values of the time lag correspond to the *broken lines*). As in the case of the ticker "NVDA" corresponding to the company "NVIDIA Corporation" in Fig. 10.2, the correlation coefficients at positive time lags are always larger than the corresponding at negative ones, this suggests that today's web search volumes anticipate and affect the trading activity of the following days (typically one or two days at most)

in a stock are reported in the news and this pushes web users to query for the corresponding ticker. In contrast, in many cases we observe the opposite, namely that query volumes tend to anticipate trading volumes. In order to prove such a non trivial features we compute the cross-correlation function of the two time series.

The cross correlation $r(\delta)$ can be viewed as the Pearson product-moment correlation coefficient between two series, with one "delayed" by a given time lag δ with respect to the second one. In particular, given two time series X_t and Y_t, the cross-correlation coefficient r at lag δ is:

$$r(\delta) = \frac{\sum_{t=1}^{n}(X_t - \overline{X})(Y_{t+\delta} - \overline{Y})}{\sqrt{\sum_{t=1}^{n}(X_t - \overline{X})^2}\sqrt{\sum_{t=1}^{n}(Y_{t+\delta} - \overline{Y})^2}} \quad (10.1)$$

where $\overline{X}$, $\overline{Y}$ are the sample average of the two time series. This cross-correlation coefficient ranges from -1 (anticorrelation) to 1 (correlation). To reduce short-term fluctuations and to reveal long-term trends, we also compute a simple moving average of width w for each time series. This smoothing procedure consists in averaging the signal at a given time t with the $w - 1$ previous points. We use $w = 1, 2, 3, 5$ ($w = 1$ being the raw series). For all the companies taken into consideration we consider $n = 250$ working days (which corresponds to one year). Hereafter we indicate the query-search volumes as X_t and the daily stock trading volumes Y_t. We compute the cross-correlation coefficient between X_t and Y_t applying a maximum lag δ of 10 working days. We assume that both processes are stationary in the whole time window composed of 250 working days. That is, we consider that the cross correlation function depends only on the time lag δ.

The lower panels of Figs. 10.2 and 10.3 report the values of cross correlation between trading and query volume as a function of the time lag δ (as defined in Eq. 10.1).

The cross correlation coefficients for positive values of δ (solid lines) are always larger than the ones for negative time lag (broken lines). This means that query volumes tend to anticipate trading volumes. Such an anticipation spans from 1 to 3 days at most.

Beyond a lag of 3 days in advance, the correlation of query volumes with trading volumes vanishes. The result is robust to varying values of the width w of the moving average window.

In order to assess the statistical significance of the results for the NASDAQ-100 set, we constructed a reshuffled data set in which the query volume time series of a company C_i is randomly paired to the trading volume time series of another company C_j. The values of the cross-correlation coefficient averaged across 1000 such permutations (values which span the range $[-0.033, 0.06]$) are smaller than the original one (which is 0.31) by a factor 10. The residual correlation present in the reshuffled dataset can be explained in terms of general trends of the market and of the specific (technological) sector considered [11].

As an additional measure of the anticipation effect, we also performed the Granger causality test [12], in order to determine if today query-log signals provide significant information on forecasting tomorrow trading volumes. We find that trading volumes can be considered Granger-caused by the query volume.

In detail the test is performed for all the 100 tickers of the considered stocks with a lag of 1 day: 39 % of trade volumes are Granger-caused by query volumes at $p = .05$; 20% of trade volumes are Granger-caused by query volumes at $p = .01$. The (macro-)average reduction in RSS for the linear autoregression models was 4.37 %. On the other hand 15 % of query volumes are Granger-caused by trade volumes at $p = .05$; 5 % of query volumes are Granger-caused by trade volumes at $p = .01$ The (macro-)average reduction in RSS for the linear autoregression models was 1.7 %. The percentages are on a per-ticker basis, i.e. the whole time series is Granger-caused or not at all. If restrict to the 87 *clean* tickers, we get the following results for a lag of 1 day: 45.35 % of trade volumes are Granger-caused by query volumes at $p = .05$, while 33.72 % of trade volumes are Granger-caused by query volumes at $p = .01$. The (macro-)average reduction in RSS for the linear autoregression models was 4.9 %. If we consider the reverse direction, . On the other hand 17.4 % of query volumes are Granger-caused by trade volumes at $p = .05$; 5.8 % of query volumes are Granger-caused by trade volumes at $p = .01$ The (macro-)average reduction in RSS for the linear autoregression models was 1.78 %. We also used the Granger causality test to study the relation between web user volumes and trade volumes. We first conducted the test on the whole set of 100 tickers, obtaining the following results for a lag of 1 day: 35 % of trade volumes are Granger-caused by query volumes at $p = .05$, while 25 % of trade-volumes are Granger-caused by the correspondent query-volume at $p = .01$. The average reduction in RSS for the linear autoregression models was 3.55 %. We then repeated the test on the 87 clean tickers. For a lag of 1 day, 40.7 % of the trading volumes are Granger-caused by the query volume at $p = .05$; 29.1 % of the trading volumes are Granger-caused by the query volume at $p = .01$. The average reduction in RSS was 4 %.

10.2 Users' Behavior

Let us now move to the second aspect which can be analyzed through the Yahoo search engine's database: the activity of single users. In fact we are able to track the users who have registered to Yahoo! and thus have a Yahoo! profile. One could expect that users regularly query a set of tickers corresponding to stocks of their interest. This is because, when querying for the ticker, the search engine returns the user, as top result of the search, a direct display of information on the trend of the stock price. In addition, if any important news appears, the corresponding page would show among the next top links in the search result. Therefore, we first compute the distribution of the number of tickers searched by each user in various time windows and time resolution (see Fig. 10.4). Interestingly, most users search only one ticker, not only within a month, but also within the whole year. This result is robust along the time interval under observation and across tickers. As a further step, among the users who search at least once a given ticker in a certain time window, we computed

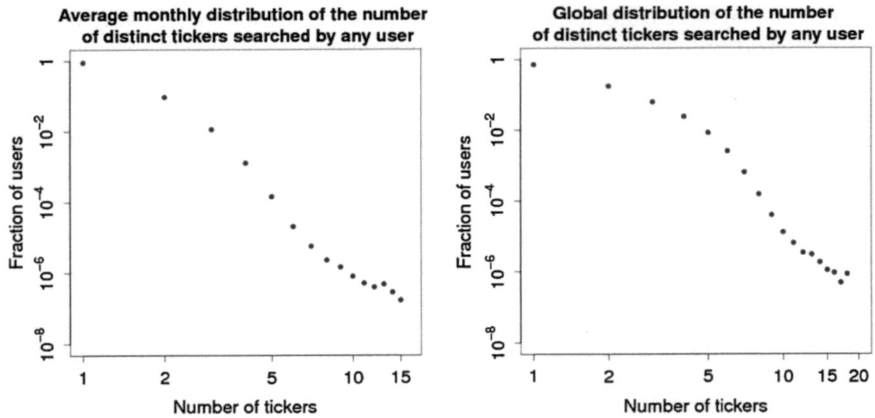

Fig. 10.4 Average (*left*) monthly and (*right*) yearly distribution of the number of distinct tickers searched by any Yahoo! user

the distribution of the number of different days in which they search again for the same ticker. In this case, we restrict the analysis to some specific tickers, namely to those with highest cross-correlation between query volumes and trading volumes (e.g., those for Apple Inc., Amazon.com, Netflix Inc.). Surprisingly, as shown in Figs. 10.5, 10.6 and 10.7, the majority of users (90 %) searched the ticker only once, not only during a month, but also within a year. Again, this result is robust along the 12 months in our dataset. Altogether, we find that most users search for one "favorite" stock, only once. The fact that these users do not check regularly a wide portfolio of stocks suggests that they are not financial experts. In addition, there is no consistent pattern over time and users perform their search in a seemingly uniform way over the months.

Overall, combining the evidence on relation between query and trading volumes with the evidence on individual user behavior brings about a quite surprising picture. Movements in trading volume can be anticipated by volumes of queries submitted by non-expert users, a sort of *wisdom of crowds* effect.

10.3 Summary and Perspectives

We crawled the information stored in query-logs of the Yahoo! search engine to assess whether signals in querying activity of web users interested in particular stocks can anticipate movements in trading activity of the same stocks. Volumes of queries for tickers of stocks are then compared with the effective trading volume of the same stocks by computing time-delayed cross-correlation.

Our results show the existence of a positive correlation between what stocks people search today on the Web and the trading volumes of those stocks in the

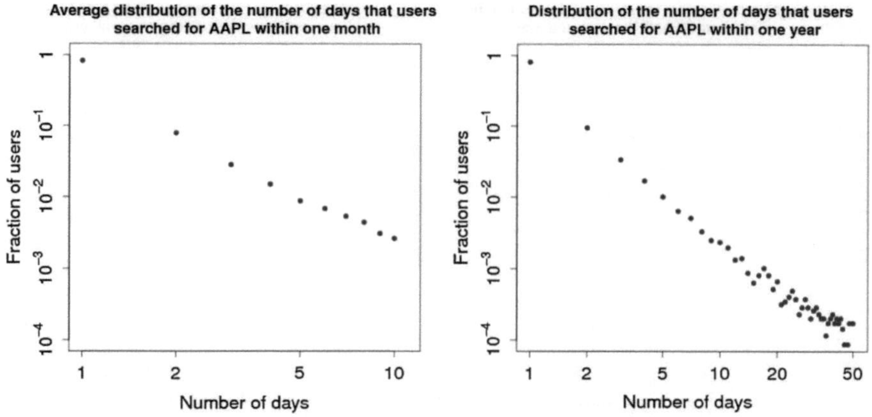

Fig. 10.5 Distribution of the number of days that users searched for AAPL within one month (*left*) and over the whole year (*right*)

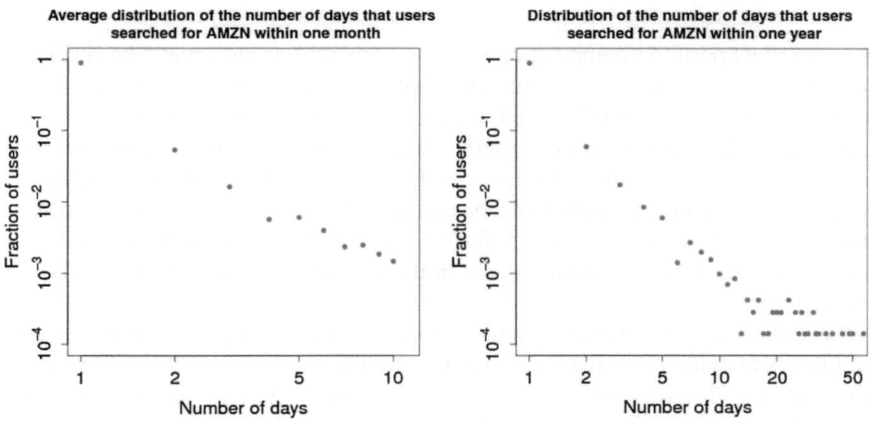

Fig. 10.6 Distribution of the number of days that users searched for AMZN within one month (*left*) and over the whole year (*right*)

following days. The direction of the correlation is confirmed by the statistical test of Granger-causality.

Furthermore, the analysis of individual users' behavior shows that most of the users query only one stock and only once in a month. This seems to suggest that movements in the market are anticipated by a sort of "wisdom of crowd" [13]. These findings do not explain the origin of the market movements but shows that that query-log volume can be a good proxy for them.

Furthermore, if one could assume that queries of a user reflect the composition of her investment portfolio, our finding would suggest that most of the investors place their investments in only one or two financial instruments. The assumption that

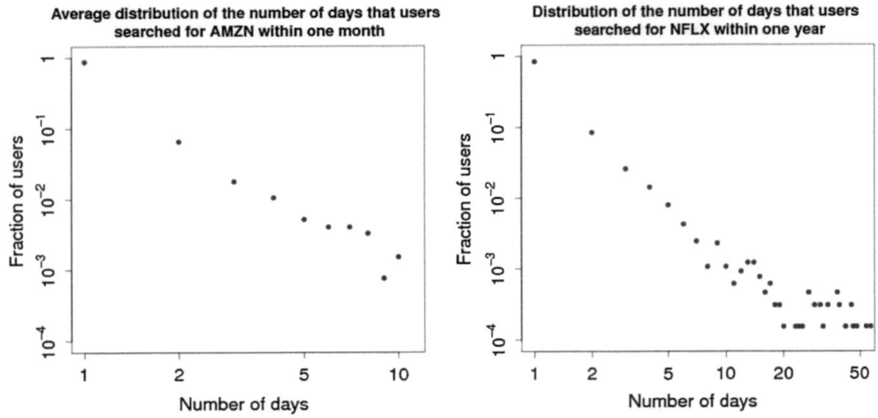

Fig. 10.7 Distribution of the number of days that users searched for NFLX within one month (*left*) and over the whole year (*right*)

queries reflect portfolio composition is a strong hypothesis and cannot be verified in our data at the current stage. The finding would then deviate from the diversification strategy of the well-known Markovitz approach, but would be in line with previous empirical works on carried out on specific financial markets. This result, if confirmed, could have very important consequences. In epidemics, by taking for granted that everybody has a mean number of contacts brings to incorrect results on disease propagations. Here the assumption that investors portfolio is balanced, while it is not, could explain why domino effects in the market are faster and more frequent than expected.

Nevertheless we want to also point out that we caution to apply straightforwardly the models of epidemic spreading to financial markets. In the latter case (differently from ordinary diseases) panic spreads mostly by news. In an ideal market, all the financial agents can become "affected" at the same time by the same piece of information. This fundamental difference makes the typical time scale of reactions in financial markets much shorter than the one in disease spreading. It is exactly for that reason that any early sign of market behavior must be considered carefully in order to promptly take the necessary countermeasures.

References

1. Bordino, I., Battiston, S., Caldarelli, G., Cristelli, M., Ukkonen, A., & Weber, I. (2012). Web search queries can predict stock market volumes. *PLoS ONE*, *7*(7), e40014. doi:10.1371/journal.pone.0040014
2. Golub, T. (2010). Counterpoint: data first. *Nature*, *464*, 679.
3. Evans, J., & Rzhetsky, A. (2010). Machine science. *Science*, *329*, 399.

4. Lazer, D., et al. (2009). Life in the network: the coming age of computational social science. *Science*, *323*, 5915.
5. Gonzalez, M., Hidalgo, C., & Barabasi, A.-L. (2008). Understanding individual human mobility patterns. *Nature*, *453*, 479.
6. Choi, H., & Varian, H. (2009). Predicting the present with google trends. Technical report
7. Goel, S., Hofman, J., Lahaie, S., Pennock, D., & Watts, D. (2010). Predicting consumer behaviour with web search. *Proceedings of the National Academy of Sciences U S A*, *107*, 17486.
8. Ginzberg, J., Mohebi, M., Patel, R., Brammer, L., Smolinski, M., & Brilliant, L. (2009). Detecting influenza epidemics using search engine query data. *Nature*, *457*, 1012.
9. Preis, T., Reith, D., & Stanley, H. (2010). Complex dynamics of our economic life on different scales: insights from search engine query data. *Philosophical Transactions of the Royal Society of London. Series A*, *368*, 5707.
10. Bollen, J., Mao, H., & Zeng, X.-J. (2011). *Journal of Computational Science*, *2*, 1.
11. Bonanno, G., Caldarelli, G., Lillo, F., Micciche, S., Vandewalle, N., & Mantegna, R. N. (2004). Networks of equities in financial markets. *EPJ B*, *38*, 363.
12. Granger, C. (1969). Investigating causal relations by econometric models and cross-spectral methods. *Econometrica*, *37*, 424.
13. Easley, D., & Kleinberg, J. (2010). *Networks, Crowds, and Markets: Reasoning About a Highly Connected World*. Cambridge: Cambridge University Press.

Part III
A Metrics for the Economic Complexity

Chapter 11
Theory of Economic Growth

11.1 Adam Smith's Theory and Specialization of the Productive System

The economic theories developed in the last two centuries to explain country growth are mainly influenced by the paradigm introduced by Adam Smith theory [1]. According to Adam Smith [1] the wealth and the richness of a country are linked to division of labour. We can say that he supported the idea that the economic efficiency is proportional to the specialization. The more a country is specialized in a productive sector, the more this country can be efficient in that production. The consequence of such an equation between specialization and efficiency is that the development and the total wealth of countries increase as the the specialization increases. However, the degree of specialization that can be reached by the global market is intimately limited by the the size of the market itself. The second consequence of Adam Smith's theory is that market with increasing size would produce an increase in the total wealth of countries since a deeper degree of specialization is reachable. In Adam Smith's theory the increasing wealth of the nations derives from the emergence of new activities (specialization) and the interaction between them at all scales. Both ingredients increase the economic complexity of the productive system of countries and according to Adam Smith the nations' wealth too.

11.1.1 Evidences Against Specialization

In the last century the degree of market globalization has always raised. The perfect global market corresponds, in Adam Smith's theory, to the largest possible market and consequently perfectly globalized markets can produce their highest degree of specialization and therefore the maximum production of global wealth. Reaching the complete globalization means that the network in which the new activities arisen from specialization is substantially fully connected and the division of labour can

M. Cristelli, *Complexity in Financial Markets*,
Springer Theses, DOI: 10.1007/978-3-319-00723-6_11,

be exploited on a global scale. Globalized markets allow for the connection of all specialized productions. As a consequence of the Adam Smith's theory, we should observe a general growth of the wealth of nations as the degree of globalization rises. Instead in the last two centuries the distance of the Gross Domestic Product (hereafter GDP) per capita has always increased and the speed of growth of this distance has accelerated (see Fig. 11.1). Following the way of reasoning of Adam Smith, his theory would predict an almost diagonal export matrix (a matrix $C \times P$ where C is the number of countries and P is the number of products, therefore the elements of this matrix specify if a country exports a specific product). Direct inspection of this matrix (see Fig. 11.2) reveals a completely different structure. The export matrix is almost triangular, this means there exist countries which have a very diversified productive structure and other ones specialized on very few products. Differently from what expected in fact those specialized countries are very poor and their productive

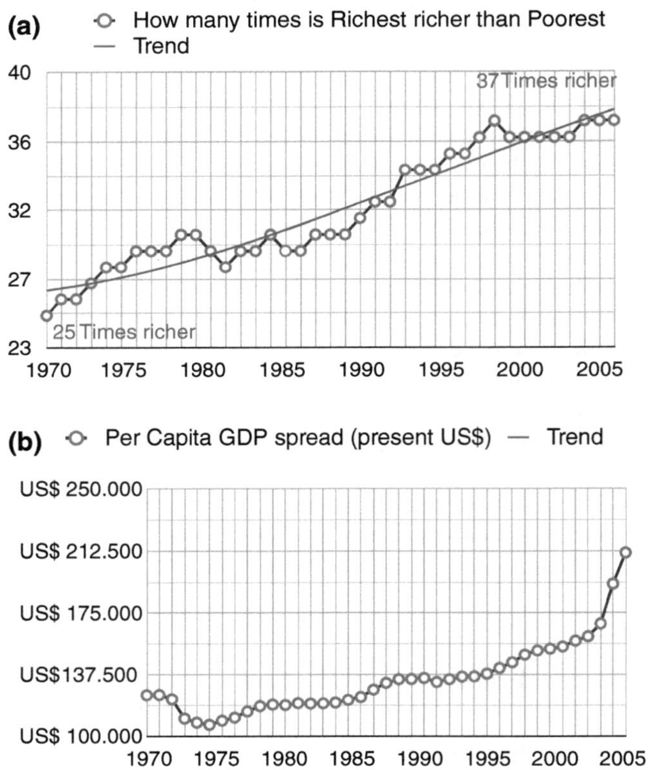

Fig. 11.1 In the last two centuries the spread of the nation wealth has become larger and larger and the speed of the divergence is getting higher and higher. This evidence contrasts Adam Smith's theories since the globalization should instead better the specialization and then the richness globally. Panel **a** evolution of how many times is the richer country respect to the poorest one; panel **b** evolution of the spread between GPD per capita

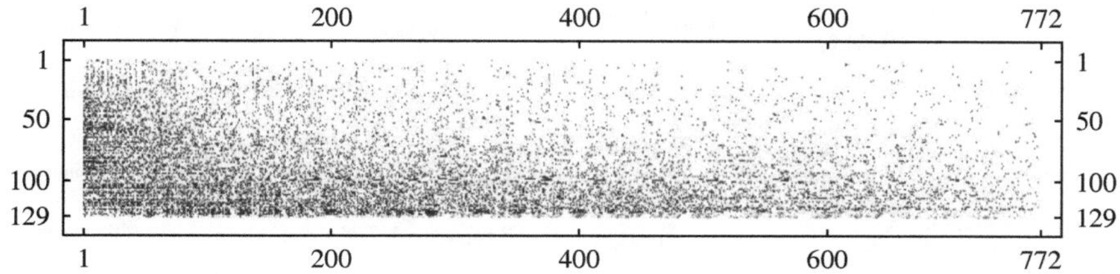

Fig. 11.2 Adam Smith's theory and most models of growth economy predict a block diagonal export matrix country-product because countries in order to improve their efficiency and therefore their wealth should specialize their productive system. Instead the empirical matrix country-product has a triangular structure revealing that the most competitive and the richest countries have a very diversified productive system able to produce almost all products, differently from Adam Smith's predictions where rich countries are highly specialized on few high quality products

system is focused on few quality products. Instead the very diversified countries are the richest ones. Then we find that empirical evidences lead us to the opposite conclusion with respect to Adam Smith's predictions: the most competitive countries with highest wealth produce almost everything and do not specialize their productive system, on the contrary they tend to diversify the production. Another prediction in this line is that neighbor countries or countries belonging to the same area should specialize their production on very different products in order to minimize competition. Instead a striking result is that neighbor countries has usually a productive basket very similar and when we group countries with respect to how the products they export are similar we usually find that countries from the same geographical area are clustered together as shown in Fig. 11.3.

It is worth noticing that in principle if all the *expertise*, labour and activities that give rise to the complexity of economic systems would be tradable and exportable then Adam Smith would be right. In such a scenario specialization would be indeed the best strategy. One possible answer to the discrepancy between specialization and diversification as the optimal growth strategy is that some activities arisen from the division of labour cannot be exported or cannot be traded. Therefore the specialization deriving from these skills are localized in a specific country/region and cannot be exploited by the global productive system. If these emerging skills are not tradable or exchangeable by consequence a country is no more able to import such skills or such specialization from another one and must develop them breaking down the principle of specialization of the productive system. Since a country cannot buy or acquire these activities in the global market, the only possible strategy of a country which needs them must be the development of them. Then in such a scenario the specialization cannot be an optimal growth strategy, on the contrary a competitive country is a country which owns the maximum number of these non importable activities. In summary if these non importable/non tradable activities exist then the observation of diversified productive systems can be interpreted as the development of all these *special* activities.

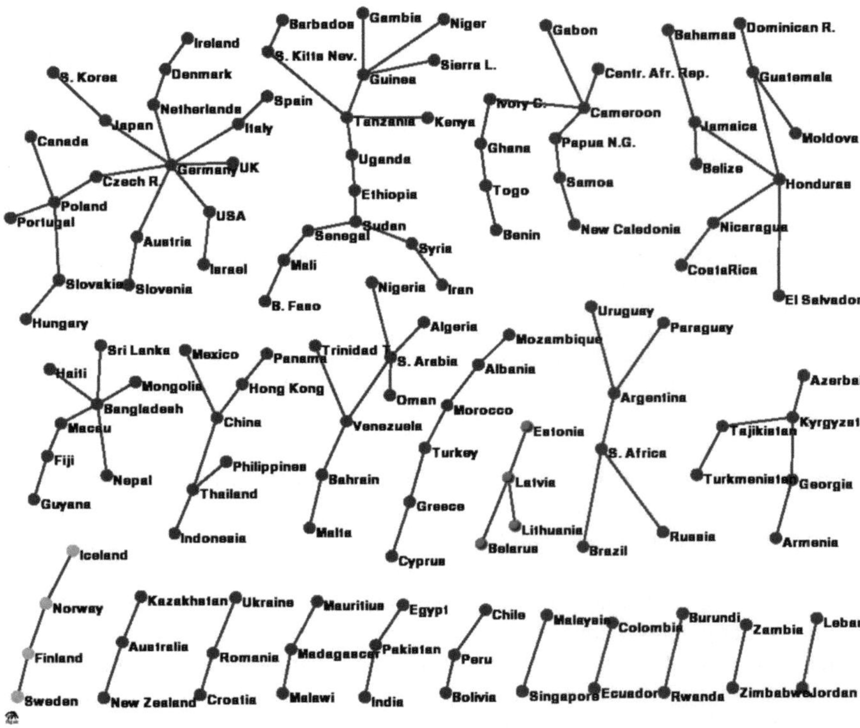

Fig. 11.3 Classical theories of economic growth also predict that neighbor countries should have productive system focused on different products in order to minimize the competition and increase the efficiency of the production. However when the similarity of the basket of products exported by countries is investigated, we find that countries belonging to the same geographical area tend to be clustered in the same group in contrast with the theoretical prediction

We are going to develop this new paradigm in next two chapters presenting two methods to measure the difference in economic complexity (i.e. the value represented by these non tradable activities) of countries while in the next section we briefly report the mainstream of growth economic theory.

11.2 Classical Explanation of Countries' Patterns of Specialization

Current growth[1] theory mainly belong to two mainstreams to explain countries' patterns of specialization.

[1] This section is mostly extracted from [2].

- The first approach is based on the country endowments of productive factors such as labor, skills, human capital, infrastructure, etc and on the proportions in which these factors are needed to produce different goods (see [3]). The consequent of such a vision is that poor countries specialize in goods that are relatively intensive in labor and land, while richer countries specialize in goods that use more human and physical capital and demand better infrastructure and institutions. Furthermore according to these models, the speed at which the factor (physical capital, say, or skills) is accumulated ultimately determines the change in the type of product the country chooses to export. The underlying assumption is that there always exists some combination of goods through which these factors can be expressed. Thus, controlling for initial factor endowments, the particular products a country produces carry no consequence for future economic performance.
- The second approach instead points out that the key element is represented by technological differences [4] and therefore needs to be complemented with a theory of what may lie behind these differences and how they may evolve over time. The two dominant theories the varieties model [4] and the quality ladders [5, 6] assume a continuum of products in some technological space. According to this line of reasoning, there is always a slightly more advanced product that countries can move to as they upgrade their technology. The world of products is abstracted away and ignored when thinking about structural transformation and growth.The abstraction from the space of products in standard economic theory is not an act of naŢvetŐ, but a natural consequence of the lack of tools available to describe them. In the next two chapter we present a method which instead introduce a metric in order to measure of the complexity of a productive system and the quality (in terms of complexity of activities say labor required) of products.

In summary it appears that economic mainstream introduces arbitrary assumptions about products' space and products' structure because there are not any adequate tool to measure the economic complexity. This line of modeling of Economics is very similar to the one seen for financial markets where classical models completely neglect Stylized Facts and the enormous discrepancy between model predictions and real market observations. In the context of economic growth the classical theory usually neglects the question whether or not the type of product a country exports matter for subsequent economic performances. The seminal texts of development economics held that it does matter, suggesting that industrialization creates externalities that lead to accelerated growth [7–9] while mainstream does not incorporated these ideas in the two families of models presented above.

References

1. Smith, A. (1776). *The wealth of nations*. London: W. Strahan and T. Cadell.
2. Hidaldo, C. A., & Hausmann, R. (2008). A network view of economic development. *Developing Alternatives*, *12*(2008), 5.
3. Flam, H., & Flanders, M. (1991). *Heckscher-Ohlin trade theory*. Cambridge, Mass.: MIT Press.

4. Romer, P. (1990). Endogenous technological change. *Journal of Political Economy*, *98*, 71.
5. Aghion, P., & Howitt, P. (1992). Endogenous technological change. *Econometrica*, *60*, 323.
6. Grossman, G. M., & Helpman, E. (1990). Quality ladders in the theory of growth. *Review of Economic Studies*, *98*, 71.
7. Rosenstein-Rodan, P. (1943). Problems of industrialization of eastern and south eastern europe. *Economic Journal*, *53*, 202.
8. Hirschman, A. O. (1958). *The strategy of economic development*. New Haven: Yale University Press.
9. Matsuyama, K. (1992). Agricultural productivity, comparative advantage, and economic growth. *Journal of Economic Theory*, *58*, 317.

Chapter 12
Beyond Classical Theory of Economic Growth

A possible answer to the puzzling discrepancy between the Adam Smith's optimal growth strategy of specialization and what is really observed (i.e. the richest countries are also the one with the most diversified productive system) can be found in the existence of non exportable (or non importable) activities which, once developed by a country or an area, stay localized in that area. In such a scenario it becomes a priority to develop all these non tradable and localized activities for a countries and the consequence of this strategy of development leads to a diversified productive system.

The difference between empirical evidences and specialization is a longstanding known fact but this new economic vision of the country growth in terms of non exchangeable activities has been very recently developed in [1].

In fact the authors of [1] proposes a theory to explain the features of the adjacency matrix M (triangular and not diagonal) whose entries M_{cp} are 1 if the country c produces the product p, 0 otherwise (as we argue in the next section the definition of the matrix M is not univocal). This matrix can be also seen as a bipartite network countries-products, see Fig. 12.1. In [1] Hausmann et al. argue that these non tradable and localized activities, called capabilities, are the key element to grasp the complexity of a productive system. Each product requires a specific set of capabilities in order to be able to make it and then to export it.

The capabilities can be identified with the presence in a country of infrastructures such as railways, airports, with an efficient legal system, with the existence of laws which favor investments, with the existence of a good educational system, with favorable climate conditions, etc. In the framework of this theory, the bipartite countries-products network is therefore just the projection of a tripartite network where capabilities are the intermediate level which links countries to products, a country can produce a product if it owns all the capabilities required (see Fig. 12.2). It is worth noticing that the level of capabilities is substantially unobservable because it is impossible to list, for real, all the capabilities required for a product because products are the result of a complex synergy among capabilities and non localized activities. It is also impossible to determine all the capabilities of a country and it

M. Cristelli, *Complexity in Financial Markets*,
Springer Theses, DOI: 10.1007/978-3-319-00723-6_12,

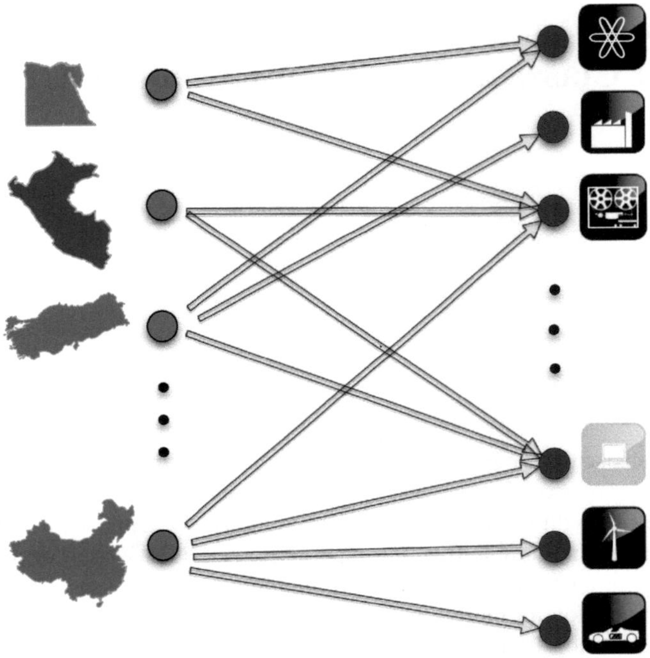

Fig. 12.1 Illustration of the bipartite network countries-products which can be also described in terms of a rectangular adjacency matrix M

is even more difficult to try to quantitatively characterize them. Thus the tripartite graph is an unobservable object but this interpretation plays an important conceptual role. The (measurable) bipartite graph is now a proxy for the complexity of the productive system by the export basket of a country because exported products give an information about the capabilities owned by a country. In this chapter we report the method developed in [1] to construct a measure for the complexity of a country (i.e. endowment of capabilities) starting from the bipartite graph (i.e. the matrix M).

12.1 Method of Reflections

The method of reflections developed in [1] introduces two variables, the former, k_c, will measure the competitiveness of a country in terms of the diversification of the production while the latter, k_p, will be a proxy for the ugliness of a product in terms of the ubiquity. By ugliness of a product we somehow intend the inverse of the complexity of a product with respect to the capabilities required, in fact if many (especially poor) countries are able to produce it we conclude (and the method must do the same) that this product cannot be complex. In such a framework, the more a

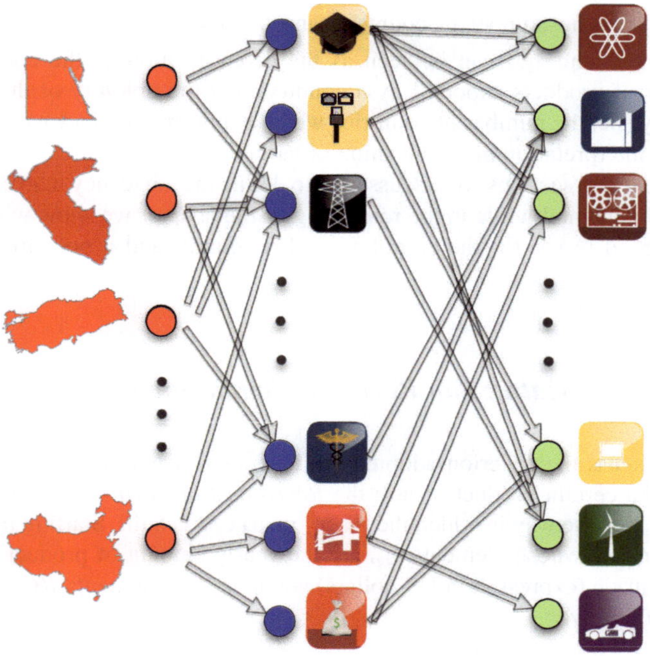

Fig. 12.2 Illustration of the tripartite network countries-capabilities-products. A country is able to make a product only if it owns all the capabilities required by that product. The intermediate level represented by the capabilities is not measurable, we can access only the projected network countries-products. However, the export basket of a country can be a proxy for the endowment of capabilities of a country which are the key determining the competitiveness and complexity of a productive system

product is ubiquitous, the more its ugliness will be high. From a mathematical point of view the method of reflection is defined iteratively as

$$\begin{aligned} k_c^{(n)} &= \frac{1}{k_c^{(0)}} \sum_{p=1}^{N_p} M_{cp}\, k_p^{(n-1)} \\ k_p^{(n)} &= \frac{1}{k_p^{(0)}} \sum_{c=1}^{N_c} M_{cp}\, k_c^{(n-1)} \end{aligned} \tag{12.1}$$

where $k_c^{(0)}$ and $k_p^{(0)}$ are the initial conditions of the iteration and they read as

$$\begin{aligned} k_c^{(0)} &= \sum_{p=1}^{N_p} M_{cp} \\ k_p^{(0)} &= \sum_{c=1}^{N_c} M_{cp}. \end{aligned} \tag{12.2}$$

N_p and N_c are the number of countries and of products respectively. The initial conditions $k_c^{(0)}$ and $k_p^{(0)}$ can be seen as the diversification of the country c (i.e. the number of products exported by a country) and the ubiquity of the product p respectively (i.e. the number of countries which export the product p). In Sect. 12.3 we give an interpretation of the meaning of variables for $n > 0$.

In the next two sections we discuss how to define the adjacency matrix M and the results of the method while in the remaining of the chapter we show why Eqs. 12.1 are not a good proxy for the complexity of a country and a poor metric for the capabilities.

12.1.1 Binary Matrix and Revealed Comparative Advantage

Here we discuss the criterion adopted in [1] to establish whether a country is an exporter of a certain product or not: the *Revealed Comparative Advantage* (*RCA*) [2]. This is then the way in which the binary matrix M is built. Starting from the raw export matrix Q whose elements q_{cp} are the yearly amount of product p exported by the country c (expressed in US dollars) we define the elements RCA_{cp} of a new matrix as it follows

$$RCA_{cp} = \frac{\dfrac{q_{cp}}{\sum_{p'} q_{cp'}}}{\dfrac{\sum_{c'} q_{c'p}}{\sum_{c'p'} q_{c'p'}}}. \tag{12.3}$$

The numerator of RCA_{cp} measures how important is the product p in the export basket of the country c, while the denominator specifies the share of the global production of the product p with respect to the whole world production (all countries and all goods). Therefore if the ratio is larger than 1 it means that the share of the product p for the economy of the country c is higher than the average share of this product in the world economy, vice-versa if it is smaller than 1. According to this observation we say that a country is an exporter of p only if it has a revealed comparative advantage in the export of this product, in formula a country is a exporter of the product p if $RCA_{cp} > 1$. Consequently the (intensive) binary matrix M is defined as

$$M_{cp} = \begin{cases} 1 & \text{if } RCA_{cp} \geq 1 \\ 0 & \text{if } RCA_{cp} < 1. \end{cases} \tag{12.4}$$

We observe the choice of the definition of the binary matrix is arbitrary, however the RCA is a commonly used and reliable measure of the export of a country [2]. In addition it is argued in [1] that around the chosen threshold to define the binary matrix a sort of transition is observed, for higher values of the threshold the elements M_{cp} are zero for almost all (c, p) while the triangular structure of M_{cp} emerges when the threshold approaches 1.

12.1.2 Main Results

We now show that the method of reflections gives a good insight on the traditional monetary indicators of country economy such as the the level of income, the GDP, etc while in Sect. 12.3 we show that the variables k_c and k_p are a poor metric for the level of complexity of a country. Therefore our task will be the development of a method which is still able to be informative on the traditional monetary indicators but also consistent with the capabilities level of a country. More details about the results of this method can be found in [1] and the Supporting Information of this paper.

12.1.2.1 Evidences for Capabilities

Let us consider the scatter plot in the plane defined by $k_c^{(0)}$ and $k_c^{(1)}$. and reported in Fig. 12.3. We recall that $k_c^{(0)}$ is the diversification of the productive system of a country while $k_c^{(1)}$ specifies how ubiquitous are the products exported by the country c. As we can see countries tend to be clustered in the first and third sector. The orthogonal solid lines represent the average of $k_c^{(0)}$ and $k_c^{(1)}$, therefore countries in the first sector are more diversified than the average and produce products which are less ubiquitous

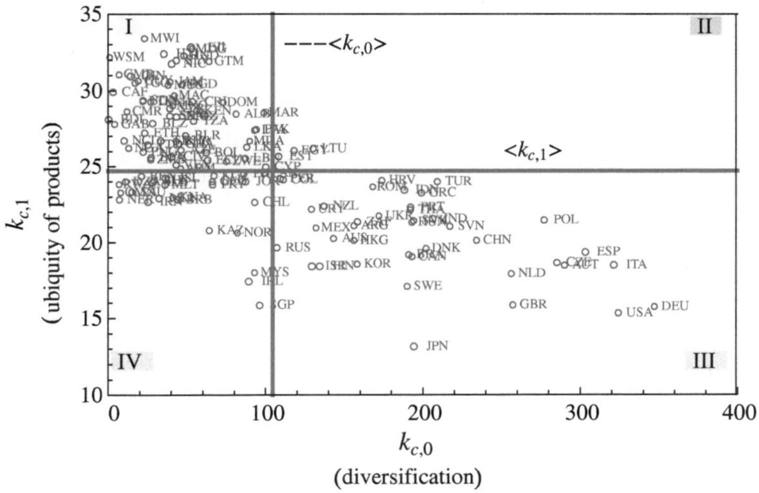

Fig. 12.3 Scatter plot in the plane defined by $k_c^{(0)}$ and $k_c^{(1)}$. The variable $k_c^{(0)}$ is the diversification of the productive system of country while $k_c^{(1)}$ specifies how ubiquitous are the products exported by the country c. As we can see countries tend to be clustered in the first and third sector. This plot can be seen as an evidence for a third and intermediate layer between countries and products because we observe a correlation among the quality of the products and the diversification of the countries

than the average of the average ubiquity, that is products which are more complex. The opposite consideration is valid for the third sector, while the two remaining sectors are almost empty. This scatter plot can be seen as an evidence for a third and intermediate layer between countries and products because we observe a correlation among the quality of the products and the diversification of the countries. The countries in the second and fourth sector can be seen as countries which are *out of equilibrium* with respect to their capabilities, in the fourth sector we may find emerging countries, in the second one countries which live above the effective competitiveness.

12.1.2.2 Correlation with the GDP

The method of reflections, as defined in Eq. 12.1 produces a shrinkage of the information as shown in Fig. 12.4. We will extensively discuss this point in Sect. 12.2 and 12.3. Therefore the authors of [1], once chosen the iteration order n at which the method is stopped, rescale the almost degenerate variables subtracting their average and dividing them by the square root of their variance. In Sect. 12.2 we give a mathematical interpretation of this procedure, in this section we just consider it as a recipe in order to extract a realistic dispersion of the variable defined by the iterative method. Given this prescription, we can study how the correlation between the variable $k_c^{(n)}$ (for n even) and the GDP per capita evolves varying the iteration order. We can see in Fig. 12.5 (left panel) that the ranking (actually we are not looking at the cardinal properties deriving from the metric defined by the variable $k_c^{(n)}$ but only to

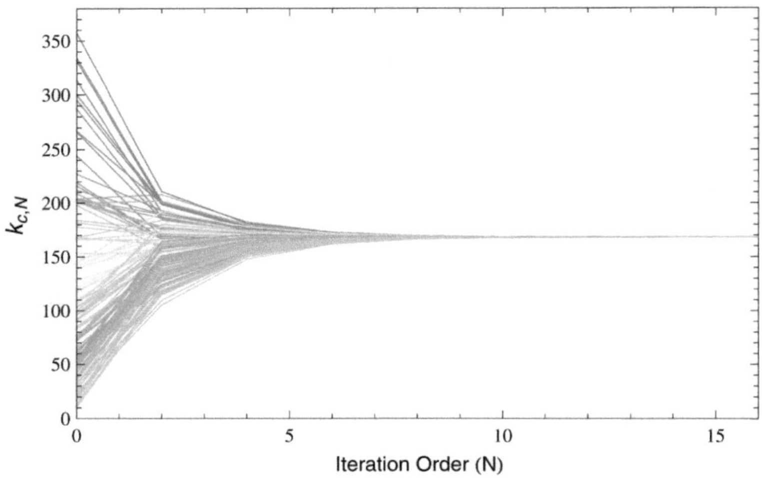

Fig. 12.4 As defined by Eqs. 12.1 the method of reflections for increasing values of the iteration order n would produce an almost degenerate metric for country complexity. Once fixed the value of n at which the method is stopped the authors of [1] rescale the variables subtracting their average and dividing them by the square root of the variance

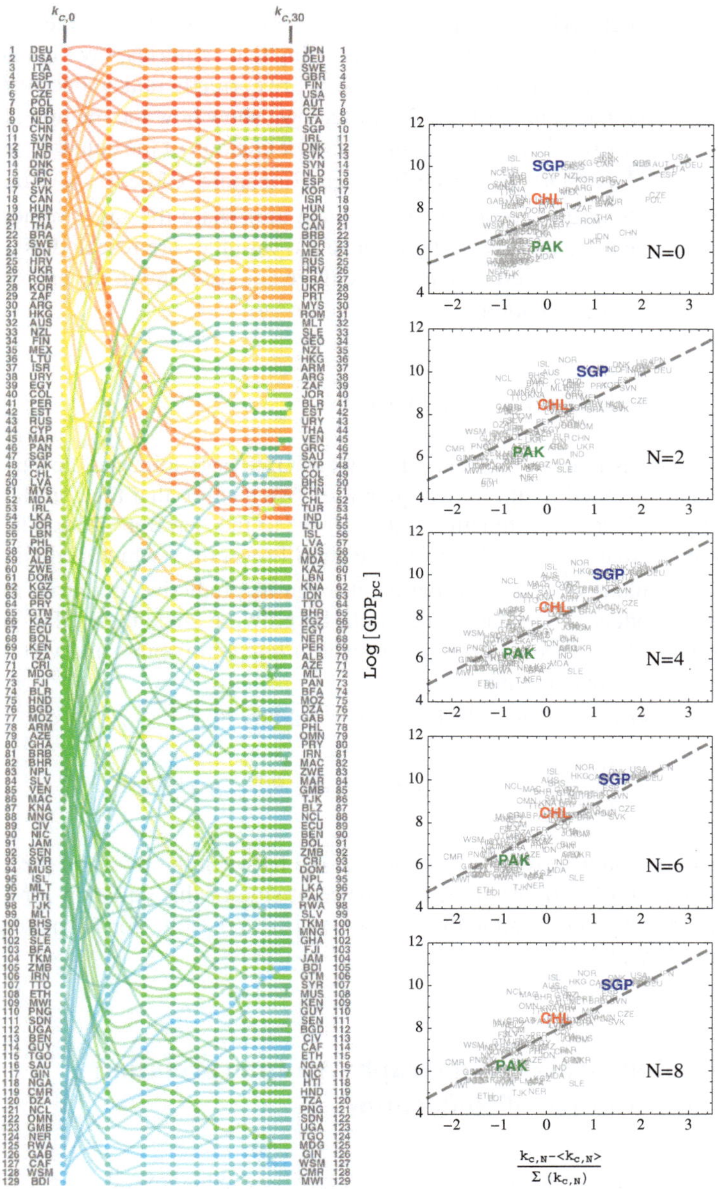

Fig. 12.5 (*Left panel*) The ranking produced by the variables $k_c^{(n)}$ (for even values of n) as a function of n is compared with the monetary ranking defined the GDP per capita. We obtain an excellent correlation among the two ranking (we are considering only an ordinal rank, we are not performing a cardinal comparison of the ranking). (*Right panel*) The iteration of the method refines the information and removes the initial degeneration of Chile, Singapore and Pakistan and ranks these three countries as expected from the GDP per capita

the ordinal properties) becomes more and more correlate with the monetary ranking defined by the GDP per capita. It is worth noticing that countries, say Finland, which start very far from their rank (in terms of GDP), are correctly ranked once the method is iterated. In the left panel of Fig. 12.5 instead we show how the method is refining the information extracted from the matrix M. The position of the countries is tracked in the plane $k_c^{(n)}$ vs GDP per capita: Pakistan, Singapore, Chile. At $n = 0$ the three countries have the same competitiveness since they share the same values of $k_c^{(0)}$. As n increases we see that the method (once the variables are rescaled as discussed above) removes the degeneration and these three countries are in the end ordered as in the GDP per capita ranking.

12.1.2.3 Prediction for Economic Growth

We can also investigate the correlation between GDP per capita and the variable $k_c^{(n)}$ by the plane specified by these two quantities. We can see that the correlation with the GDP is an increasing function with respect to the iteration order. The authors of [1] propose that the deviation from the equilibrium line (which is defined as the best fit in the plane $k_c^{(n)}$ vs GDP per capita) can be predictive, countries below the line have a productive system which could be growing since their GDP per capita is lower than what expected given their level of competitiveness and vice-versa for those countries above the line.

In detail they perform a multi-parameters regression

$$\ln \frac{GDP(t+\Delta t)}{GDP(t)} = \alpha GDP(t) + \beta_1 k_c^{(2n)} + \beta_2 k_c^{(2n+1)} + \gamma \tag{12.5}$$

The author find a correlation around 0.2–0.3 between the return of the GDP and the combination of variables given by the right side of this equation for $\Delta t = 10$ and 20 years, while correlation drops below 0.1 for $\Delta t < 10$, for further details see Supplementary Information of [1].

12.2 Random Walks on a Graph: An Interpretation of the Method of Reflections

Let us now give a mathematical interpretation of the method of reflections in terms of a random walk on a graph. We start outlining that Eqs. 12.1 must be distinguished in *even* and *odd* iterations: in the even iteration $2n$ the variables $k_c^{(2n+2)}$ can be seen in some way as a refined version (even if they change their meaning, see Sect. 12.3.1) of $k_c^{(2n)}$. In fact the variables k_c and k_p are alternatively the ubiquity or the diversification of *something*, therefore two iterations are needed to start from a certain type of variable—ubiquity or diversification—and obtain a variable of the same kind. In the case of $k_c^{(2n)}$ taking only even iterations we are considering only the variables associated to countries whose type is *diversification*. Equivalently $k_p^{(2n+2)}$ can be interpreted as an ubiquity as $k_p^{(2n)}$ (i.e. as the *dis-value* of the product p). Instead the odd iterations $k_c^{(2n+1)}$ are the arithmetic mean of the ubiquity $k_p^{(2n)}$ of the products p produced by the country c, while $k_p^{(2n+1)}$ is the arithmetic mean of the diversification $k_c^{(2n)}$ of the countries c producing p.

From the observation that we must somehow look at the method of reflections each two iterations we are going to show that the iterative equation can be interpreted as a random walk on a suitable bipartite graph country-product. As a preliminary step we introduce two complementary graphs. These are the network of countries and the network of products that we obtain from the original set of relations represented by the binary matrix M. The idea is to connect the various countries with a link whose strength is given by the number of products they mutually produce. Similarly, we connect two products if one or more countries produce them. The first and most immediate way to achieve such result is to consider the two matrix products

$$\begin{cases} C = MM^T \\ P = M^T M \end{cases} \tag{12.6}$$

which define two square matrices the *country–country* matrix C ($N_C \times N_C$) and the *product–product* matrix ($N_P \times N_P$). The element C_{ij} of the first one corresponds to the number of products that countries i and j have in common (or the total number of products produced by country i when considering the diagonal values C_{ii}). Similarly the generic element P_{ij} of the second one returns the number of countries both producing products i and j (or the total number of countries producing product i when considering the diagonal values P_{ii}).

By using the theory of stochastic or Markov matrices we show that the iterations of the method of reflection (see Eqs. 12.1) converges to a trivial fixed point. Despite such a feature, this procedure gives a ranking that becomes stable for n high enough. To such a purpose we transform the previously introduced square matrices C and P into their stochastic counterparts. Let us first define two diagonal square matrices A (matrix order N_c) and B (matrix order N_p) whose diagonal elements are the degrees k_c and k_p respectively. We can use these two matrices to define a larger (order $N_c + N_p$) matrix H with the following structure

$$H = \begin{pmatrix} 0 & A^{-1}M \\ B^{-1}M^T & 0 \end{pmatrix} \tag{12.7}$$

where M^T is the transpose of M; Eqs. 12.1 can then be rewritten in a vectorial form as

$$\begin{pmatrix} \mathbf{k_c}^{(n+1)} \\ \mathbf{k_p}^{(n+1)} \end{pmatrix} = H \begin{pmatrix} \mathbf{k_c}^{(n)} \\ \mathbf{k_p}^{(n)} \end{pmatrix} = H^{n+1} \begin{pmatrix} \mathbf{k_c}^{(0)} \\ \mathbf{k_p}^{(0)} \end{pmatrix}. \tag{12.8}$$

For n even, Eq. 12.8 separates into the following two distinct equations for the k_c s and the k_ps:

$$\begin{cases} \mathbf{k_c}^{(2n)} = \mathcal{C}\mathbf{k_c}^{(2n-2)} = \mathcal{C}^n \mathbf{k_c}^{(0)} \\ \mathbf{k_p}^{(2n)} = \mathcal{P}\mathbf{k_p}^{(2n-2)} = \mathcal{P}^n \mathbf{k_p}^{(0)} \end{cases} \tag{12.9}$$

where we have introduced the *stochastic country–country* matrix

$$\mathcal{C} = A^{-1} M B^{-1} M^T \tag{12.10}$$

and the *stochastic product–product* matrix

$$\mathcal{P} = B^{-1} M^T A^{-1} M\,. \tag{12.11}$$

Since the matrices $\mathcal{C}$ and $\mathcal{P}$ are *right stochastic* (i.e. *Markov*) *matrices*, they both satisfy the conditions:

- $\mathcal{C}_{cc'} \geq 0$ and $\mathcal{P}_{pp'} \geq 0$ for any pair (c, c') and (p, p');
- $\sum_{c'} \mathcal{C}_{cc'} = 1$ and $\sum_{p'} \mathcal{P}_{pp'} = 1$.

However it is important to stress that Eqs. 12.9 do not define two Markov chains as the matrices operate from the left with a row-column product on the column vectors $\mathbf{k_c}^{(2n)}$ and $\mathbf{k_p}^{(2n)}$, instead from right on row vectors as in the standard convention[3]. For this reason the sum of the components of the vectors $\mathbf{k_c}^{(2n)}$ and $\mathbf{k_p}^{(2n)}$ is in general non-conserved by the transformation.

If a fixed point $(\mathbf{k_c}^*, \mathbf{k_p}^*)$ exists for Eq. 12.8, i.e. Eqs. 12.9, it holds

$$\begin{cases} \mathbf{k_c}^* = \mathcal{C}\mathbf{k_c}^* \\ \mathbf{k_p}^* = \mathcal{P}\mathbf{k_p}^* \end{cases} \tag{12.12}$$

Therefore, if the two matrices are *ergodic*, the final states of the iteration correspond to the eigenvectors of the matrices C and P with unitary eigenvalue.

12.2.1 Spectral Properties

Given the previous consideration instead of considering the iterative procedure of the method of reflections we can now simply analyze the spectral properties of the matrices $\mathcal{C}$ and $\mathcal{P}$. Let us focus on the first one, since similar considerations hold for the matrix $\mathcal{P}$. As aforementioned the matrix $\mathcal{C}$ is a stochastic matrix. It follows that all the eigenvalues λ_i are bounded in modulus by one. Moreover, if $\mathcal{C}$ is ergodic, it has a single linearly independent *right* eigenvector[1] $\mathbf{e}_1$ corresponding

[1] The corresponding *left* eigenvector is the stationary (and asymptotic) Markov state of the Markov chain defined by the transition matrix $\mathcal{C}$.

to the highest eigenvalue $\lambda_1 = 1$ [4, 5]. Therefore $\mathbf{k_c}^{(2n)} \propto \mathbf{e}_1 + \mathcal{O}\left(\lambda_2^n\right)$ which exponentially approaches the fixed point $\mathcal{C}\mathbf{k_c}^* = \mathbf{k_c}^* \propto \mathbf{e}_1$ of eq. 12.9. Since $\mathcal{C}$ is an ergodic right stochastic matrix, it is simple to show that such eigenvector $\mathbf{e}_1$ is trivial and corresponds to a vector where all the components are the same $(e_1)_c = k$ independently of c.

Such a spectral approach clarifies better the origin of the ranking of different countries. Let us order the eigenvalues of C so that $1 = \lambda_1 > \lambda_2 \geq \cdots \geq \lambda_{N_C} \geq 0$; if both the first two eigenvalues are non-degenerate, one can approximate

$$\mathbf{k_c}^{(2n)} \propto a_1\mathbf{e}_1 + \lambda_2^n a_2\mathbf{e}_2 + \mathcal{O}\left(\lambda_3^n\right) \tag{12.13}$$

where $\mathbf{e}_2$ is the eigenvector related to the second eigenvalue λ_2, and a_i is the component of the initial fitness vector $\mathbf{k_c}^{(0)}$ (given by the diversification k_c of all the countries) along the eigenvector $\mathbf{e}_i$. After a finite number of iterations the "$(2n)^{th}$" fitness of a country c is $k_c^{(2n)} \sim a_1 k + \lambda_2^n a_2 (e_2)_c$: apart a constant shift k, the ranking of the cth country is therefore asymptotically fixed by the cth component $(e_2)_c$ of the second eigenvector.

An equivalent procedure is adopted in the original papers [1, 6], where the ranking of the countries at the $(2n)^{th}$ step of the algorithm is obtained by subtracting the arithmetic average of the components and by dividing by the standard deviation. The first operation basically eliminates the constant shift $a_1 k$ while the second operation keeps constant the magnitude of the rankings as standard deviation is asymptotically $\propto a_2\lambda_2^n$).

With the spectral analysis of the fixed point matrix it is therefore not necessary to fix an arbitrary number of iterations to obtain the relationship between country fitnesses and product ubiquity but we are able to assign the ranking (fitnesses) of the countries by simply calculating the components of the second eigenvector of C. Notice that, although if $\mathbf{e}_2$ is an eigenvector also $-\mathbf{e}_2$ is a valid eigenvector, the sign is fixed by the initial choice of $\mathbf{k_c}^{(0)}$.

We also notice that in an analogous way the matrix P fixes a ranking for the ubiquity of the products. Indeed the two matrices C and P share the same spectrum. This is not unexpected since the two matrices C and P are linked to MM^T and M^TM (same spectrum apart the degeneracy of the null space) via the affine transformations induced by A^{-1} and B^{-1} (affine transformations leave spectra invariant).

As a final remark of this section we highlight that the method of reflections appear to be somehow unnatural since we have shown that it corresponds to the search of the right eigenvector of a suitable stochastic matrix but from the theory of Markov chain we know that right eigenvectors are trivial while the information is contained in the left eigenvectors such as the invariant measure (in [7] instead we report the results of the investigation of the ranking deriving from the *left* spectral properties).

12.3 Criticisms to the Method of Reflections

The method of reflections has some very weak points which here are listed and discussed in the following of this section:

1. Variables change their meaning.
2. The variables k_c are not correlated with countries' capabilities.
3. Shrinkage of information when the method goes towards convergence.
4. Which is the correct asymptotic variable: $k^{(odd)}$ vs $k^{(even)}$.
5. Locality of regressions.

This method in fact does not reflect the original intention of the authors as we show the variables $k_c^{(n)}$ which should measure the complexity and the competitiveness of a country are not correlated with the endowments of capabilities of a country. Therefore we need to define a different mathematical transformation to measure the complexity of a productive system and of products as we are going to show in the next chapter. This new method must take into account all the aspects here listed in order to develop a theory which effectively realize the theory of capabilities developed by Hausmann and Hidalgo in [1].

12.3.1 Variable Meaning Depends on the Iteration

The (economic) interpretation of the meaning of the variables $k_c^{(n)}$ and $k_p^{(n)}$ depends on the order n of the iteration. It is easy to identify the meaning of the first iteration but the method of reflections does not preserve this meaning and does not simply refine the information. Instead at each step the method introduces a couple of variables which, alternately, can be seen as a generalized ubiquity and a generalized diversification. We say *generalized* since it is impossible to simply translate in words which quantity is ubiquitous and which one is diversified. As an example of this fact see Table 12.1.

12.3.2 Toy Model 1: $k_c^{(n)}$ are not Correlated with Countries' Capabilities

As we have said in the previous chapter, the aim of the method of reflections would be to obtain a set of self-consistent variables which are able to reproduce the distribution of the capabilities of countries. In the case of the real matrix M_{cp} of countries' export we cannot directly access to the vector of capabilities of a country. However, in order to test the method of reflections we can construct an artificial tripartite graph countries-capabilities-products. In this respect we define two matrices: a country-capabilities C_{ca} matrix whose entries specify which capabilities are owned by a country and a capabilities-product P_{ap} matrix which specifies which capabilities are

Table 12.1 The meaning of the variables $k_c^{(n)}$ and $k_p^{(n)}$ depends on the iteration order n

Variables	Type	Interpretation
$k_c^{(0)}$	Diversification	Number of products exported by c
$k_p^{(0)}$	Ubiquity	Number of countries exporting p
$k_c^{(1)}$	Ubiquity	How ubiquitous are the products exported by c on average
$k_p^{(1)}$	Diversification	How diversified are the countries exporting p on average
$k_c^{(2)}$	Diversification	How diversified are the countries exporting the same products of c on average
$k_p^{(2)}$	Ubiquity	How ubiquitous are the products exported by the countries which export p on average
$k_c^{(3)}$	Ubiquity	-
$k_p^{(3)}$	Diversification	-

required to make a product. Performing the product $C_{ca} \times P_{ap}$ of these two matrices we obtain a country-product matrix M_{cp} whose entries have the same meaning of the a real matrix M_{cp}. Differently from the matrix extracted from real data now we also know the number of capabilities of each country by construction. Then we can apply the iteration method here introduced to this artificially generated M_{cp} matrix and investigate how the variables $k_c^{(n)}$ of a country get correlated (or uncorrelated) with the respective number of capabilities of this country.

We define the C_{ca} and P_{ap} matrix as in citeHH2: the matrices C_{ca} and P_{ap} are random matrices. The elements of C_{ca} are 1 with probability $r = 0.7$ and 0 with probability $1 - r = 0.3$. Instead the elements of P_{ap} are equal to 1 with probability $q = 0.05$ and 0 with probability $1 - q = 0.95$. A more realistic test with a triangular C_{ca} matrix is one of our future goals in order to reflect the real structure of the M_{cp} matrix and to perform intensive investigation in a controlled set-up where the tripartite graph is known. In this artificial system we consider 200 capabilities while the number of countries and products is equal to the ones of the real matrix, i.e about 700 products and 140 countries. In Fig. 12.6 we give a visual representation of the three matrices M_{cp}, P_{ap} and C_{ca}.

In left panel of Fig. 12.7 we plot the $k_c^{(n)}$ obtained with the reflection method for different values of the iteration index n. As we can see the correlation between capabilities decreases when n increases so that we conclude that the $k_c^{(n)}$ does match the capabilities of a country. In addition odd iterations are positively correlated with the number of capabilities while even ones are negatively correlated, therefore it is not clear which are the correct asymptotic value that must be taken into account (see Sect. 12.3.4 for further details). The decreasing trend of the correlation of $k_c^{(n)}$ variables with the number of capabilities owned by a country is shown in Fig. 12.8 where the maximum correlation is observed at zero-order iteration.

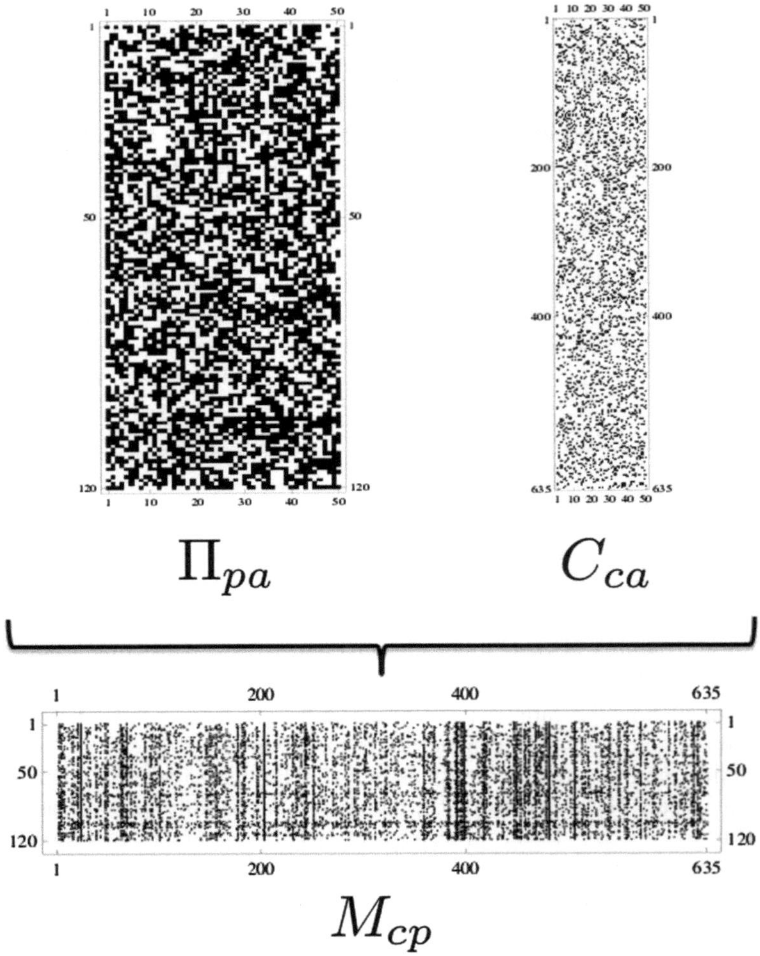

Fig. 12.6 Visual representation of M_{cp}, P_{ap} and C_{ca}. P_{ap} and C_{ca} are binary random matrix and the probability that an entry is equal to 1 is 0.05 and 0.7 respectively

12.3.3 Toy Model 2: Shrinkage of Information

Let us now suppose to consider an artificial export system composed of 7 countries and 26 products to show that the method of reflection tends to make all countries' competitiveness equal. As for the real case, we can characterize this system by introducing a binary 7×26 M_{cp} matrix in which the element M_{cp} is 1 if the country c exports the product p, according to a suitable threshold (see discussion about RCA in the previous section). We define the M_{cp} matrix of this artificial system in order to reflect the property of the real M_{cp} matrix, that is to have a triangular structure

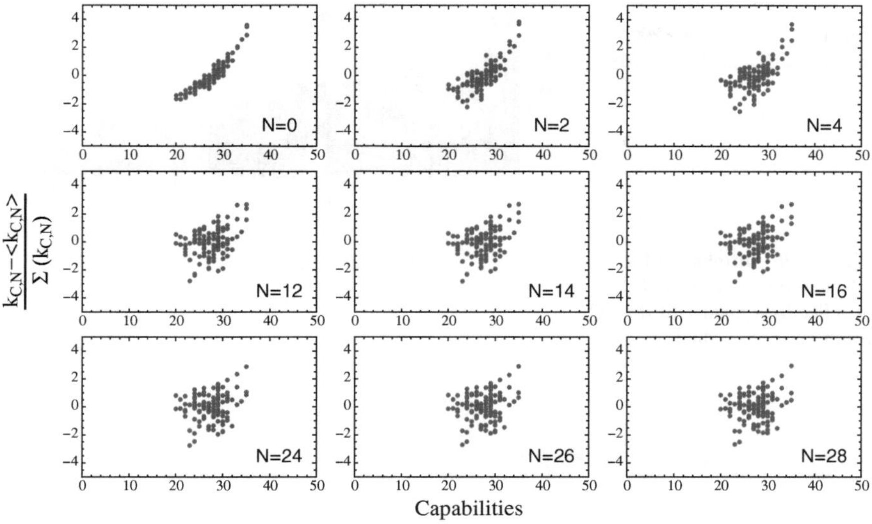

Fig. 12.7 $k_c^{(n)}$ vs capabilities: the variables introduced by the method of reflections tends to get uncorrelated with the capabilities endowments of a country. In addition odd iteration are positively correlated with capabilities while even ones are negatively correlated

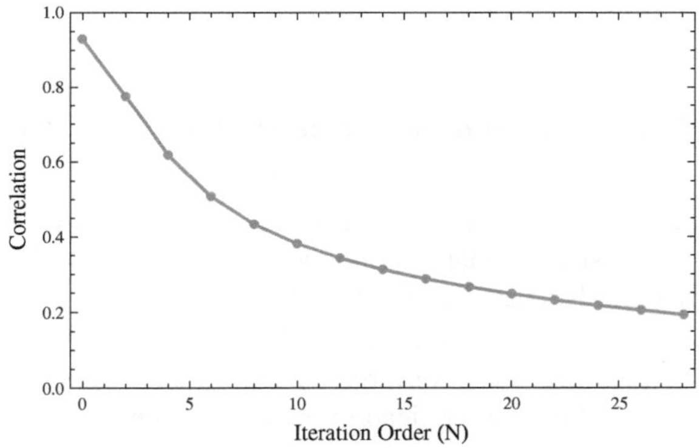

Fig. 12.8 The correlation between $k_c^{(2n)}$ and capabilities is destroyed by the method of reflection since the maximum correlation is observed for $k_c^{(0)}$, i.e. the initial diversification of a country

(see Fig. 12.9 for a graphical representation of the 7×26 matrix used). Therefore the most competitive countries of our system have a large diversification of their production and consequently they make almost all products while decreasing the competitiveness of our artificial countries, the production becomes more and more specialized on a small subset of poor quality products.

This very simple case study allows us to clearly illustrate the action of the iterations defined by the method of reflections. In Fig. 12.10 we show the results of the iterations applied to the artificial 7×26 M_{cp} matrix . The linear iteration unavoidably drives all the countries' complexity to the same value.

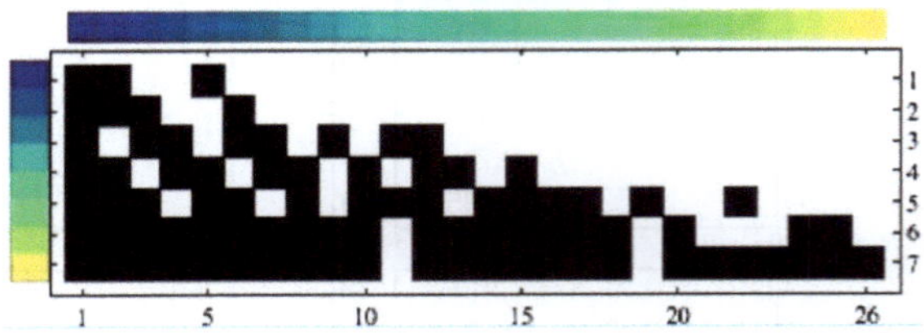

Fig. 12.9 Visual representation of the 7 × 26 M_{cp} matrix

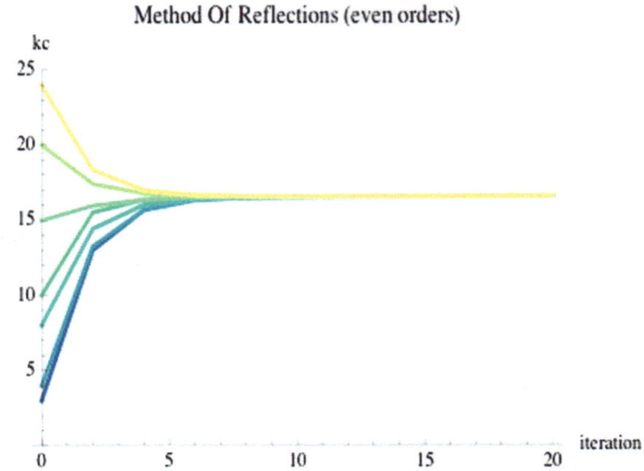

Fig. 12.10 Shrinkage of information. The method of reflections makes all the competitiveness equal

12.3.4 Asymptotic Iteration and $k^{(odd)}$ *vs* $k^{(even)}$

In [1] both the odd and the even iteration of the rescaled $k_c^{(n)}$ are said to be predictive with respect to many economic aspect such as the level of income of a certain year and the 10 years growth of the GDP of a country.

We have seen in Sect. 12.2 that the variables $k_c^{(n)}$, if not rescaled dividing them by their variance, converge all to the same values. Therefore in such contest both even and odd orders lead to the destruction of the information. When we consider the rescaled $k_c^{(n)}$ we have argued that these variables are proportional to the entries of the right second eigenvector of the stochastic matrix defined in section 12.2, the first eigenvector is a vector composed of all constants, all equal and corresponds to the non rescaled variables.

We have also shown that in the framework of this interpretation the mathematically meaningful variables are only the even iteration, that is $k_c^{(2n)}$. This fact sets two conceptual problems with respect to the method of reflections. Firstly, which are then the correct n at which the iterative system must be stopped since in the limit $n \to \infty$ we cannot consider even or odd iterations. On the other hand why the even orders $k_c^{(2n+1)}$ should be used in the predictive procedures and how they can be interpreted. In fact in the spirit of Sect. 12.2 they appear to be simply intermediate results while the relevant quantity, from a mathematical point of view, are the even orders.

In other words the method of the reflections will asymptotically oscillate between odd and even iterations and therefore the iteration procedure must be necessarily stopped at a finite order n in order to obtain the value of the variables k_p and k_c. This fact sets the crucial problem of the choice of the order n at which the procedure must be stopped, in fact this choice is now completely arbitrary differently from an asymptotic quantity because there is no more a natural candidate for a self-consistent variable.

12.3.5 Locality of Regressions

In [1] the authors perform a multi-variable regression which in formula reads as

$$\ln \frac{GDP(t + \Delta t)}{GDP(t)} = \alpha GDP(t) + \beta_1 k_c^{(2n)} + \beta_2 k_c^{(2n+1)} + \gamma \tag{12.14}$$

where α, β_1, β_2 and γ are the free parameters of the regression. Performing the regressions in this way does not really looks at the comparison of a countries with respect to the others. In fact the right side of the equation of the regression contains only local terms (the eigenvector of the country, the GDP of the country, etc) and the comparison is somehow performed only a posteriori by the optimal hyperplane of the regression or, in an equivalent way, we can say that the comparison is performed by the left side of the regression equation that is the log returns of GDP. In this sense this is an a posteriori comparison of the relative positions of countries in the space in which the regression is made. Instead as we are going to see in the next chapter we propose that the key variable of the regression must the distance from the central line in the $Fitness - GDP$ plane which is an a priori comparison of the relative positions of countries and this comparison is independent on the regression differently from the a posteriori comparison proposed in [1] that is instead regression-dependent.

12.4 Summary and Perspectives

Adam Smith's theory would predict a nearly block diagonal matrix M of countries' export, instead the existence of non tradable and localized activities which emerge with labour specialization makes diversification of production an ideal strategy for the development of a country. This means that the M matrix is a triangular-like matrix. These localized activities are called *capabilities* in [1, 6] and we can interpret the productive system of a country in terms of a tripartite graph in which capabilities are the intermediate level. In fact a country is able to produce and then to export a product p if it owns all the capabilities required by the product p. Therefore the bipartite graph countries-products specified by the matrix M is just the projection of this tripartite network. While the bipartite network is the real information which can be measured, we do not have access to the intermediate level of capabilities given their features which are almost impossible to quantify. In fact the capabilities of a country may be the educational system, the efficiency of the legal system, the presence of infrastructures, favorable climate condition, etc. Thus if we accept this interpretation in terms of an underlying inaccessible layer, we observed that the export basket of a country can be a proxy to reconstruct the endowment of capabilities of a country. The method of reflections introduced by Hausmann et al. aims at picking the information of the matrix M in order to measure the complexity (and then the competitiveness) of a country production. We have shown that this method is not able to grasp the capabilities of a country as the correlation between the output of the iterative method becomes uncorrelated with the capabilities in a toy model where we assign the tripartite graph and so we know the level of capabilities. Our aim is then to develop a new method which produces a consistent metric for the complexity of countries and the quality of products.

References

1. Hidalgo, C. A., & Hausmann, R. (2009). The building blocks of economic complexity. *Proceedings of the National Academy of Science*, *106*, 10570.
2. Balassa, B. (1965). Trade liberalization and 'revealed' comparative advantage. *Manchester School*, *33*, 99.
3. Feller, W. (1968). *An introduction to probabbility and its applications* (Vol. 1). New York: Wiley.
4. Bellman, R. (1997). *Introduction to matrix analysis* (2nd ed.). Philadelphia: Society for Industrial and Applied Mathematics.
5. Marcus, M., & Minc, H. (1964). A survey of matrix theory and matrix inequalities. Boston: Allyn & Bacon.
6. Hidalgo, C. A., Klinger, B., Barabási, A. L., & Hausmann, R. (2007). The product space conditions the development of nations. *Science*, *317*, 482.
7. Caldarelli, G., Cristelli, M., Gabrielli, A., Pietronero, L., Scala, A., and Tacchella, A. A network analysis of countries' export flows: firm grounds for the building blocks of the economy. PLoS ONE, 7 (2012), e47278. http://arxiv.org/abs/arXiv:1108.2590

Chapter 13
A New Metric for the Competitiveness of Countries and the Quality of Products

In the previous chapter we have seen that there exist some evidences for a third intermediate layer in the network countries-products. This layer is substantially unobservable but contains all the information about the complexity, the competitiveness and the growth potentiality of a productive system. The ideal picture of a tripartite network in which a country is able to export a product only if it has the capabilities required by that product highlights that, even if shrunk, the projected bipartite network *countries-products* can be informative about the layer of capabilities. In fact from a suitable analysis of the basket of exported products the endowment of capabilities may be extracted or at least measured in a consistent way. In the previous chapter we have described the attempt presented in [1]. We have seen that while the results of this method are in agreement with the traditional monetary indicators, they fail to grasp the level of complexity of a country. Here we develop a new self-consistent method which defines a metric for country competitiveness (hereafter *fitness*) and one for product qualities. On one hand we find the metric for countries matches the known results given by monetary indicators, such as GDP and GDP per capita, but, differently from the method of reflections, we are able to show that in a toy model where the tripartite graph is assigned the fitnesses of countries are highly correlated with the basket of capabilities.

13.1 A New Metric: Country Fitness and Product Quality

The theory of capabilities can be summarized saying that the more a country owns capabilities, the more its productive system is competitive and the more a product is complex, the more it requires capabilities that only complex productive systems have developed. Now it is clear why the method of reflections fails to mathematically translate this concept, in fact the quality of a product is given by the average of the competitiveness of the producers of that product. The definition of the competitiveness for countries in terms of the average of the quality of the exported products

M. Cristelli, *Complexity in Financial Markets*,
Springer Theses, DOI: 10.1007/978-3-319-00723-6_13,

is instead closer to the spirit of the capabilities framework. The definition of the product quality is problematic for the following reason. According to the theory of capabilities therefore the knowledge that a product is exported by a rich and developed country is not informative on the complexity of this product, since the most complex productive systems are completely diversified. Instead the knowledge that a product is exported by a poor country is highly informative and means that few and low level capabilities are required by this product. Equivalently, if we know that a product is exported only by complex countries then we can conclude that the product is complex. Therefore the right method to detect such a feature cannot be linear but instead a non linear algorithm is required.

Let us consider a simple example to clarify this *negative* characterization of the information about the product quality. We consider a world composed by only two countries, say USA and Nigeria, and their fitnesses (we recall that is the new name given to the variables of the metric discussed in this chapter) are, respectively, 10 and 0.1. Let us now consider two products, one made by both countries and the second made only by USA. In the method of reflections the quality of the products made by both countries will be the average of the fitnesses of USA and Nigeria, that is about 5, while the quality of the product made by the USA will be 10. We observe that USA are 100 times more competitive than Nigeria but this information is not reflected (and indeed shrunk) by the quality of the products, in fact the product made only by USA is only twice times more complex than the one made also by Nigeria, while instead the information that a country with so low fitness is able to produce it should be much more negative and the quality of this product should be of the same order of magnitude of the worst country which is able to export it. We must define a method in which the quality of the product made by both countries should be around 0.1.

For the same reason, that is the quality of a product is established by the low-ranked countries, in order to determine the value of nails we do not have to look at Germany as an exporter of them, rather we must search if, listed among the nail exporters, there are some countries of the Third or Second World. If this is the case, then nails are a low quality product. In summary, if a poor country makes a product then the level of this product cannot be high in terms of capabilities because a low competitive economy is able to produce it.

These simple examples show how a method in order to be consistent with the picture of the tripartite network must be constructed and with the informative content of the capabilities. In the next sections we describe how this philosophy can be a mathematical tool able to define a quantitative metric for country fitnesses and product qualities.

13.1.1 Intensive Case

The method described for two countries and two products is the more extremal way to implement our considerations, in fact we was assuming that the quality of the product made both by USA and Nigeria is equal to the fitness of Nigeria. This extremal method would completely ignore part of the information contained in the adjacency M and would introduce some bias in the estimation of small but highly developed countries which does not have a diversified productive system, say Finland, Austria or Switzerland. In other words, we simply want a method which gives a larger weight to the less competitive exporters of a given product. This can be achieved in formula in the following way (the method here defined is still very extremal but not as much as the previous one). The iterative methods is composed of two steps at each iteration, we first compute the intermediate variables $\tilde{F}_c^{(n)}$ and $\tilde{Q}_p^{(n)}$

$$\tilde{F}_c^{(n)} = \sum_p M_{cp} Q_p^{(n-1)} \tag{13.1}$$

$$\tilde{Q}_p^{(n)} = \frac{1}{\sum_c M_{cp} \frac{1}{F_c^{(n-1)}}} \tag{13.2}$$

and then we define the country fitness and the product quality at the order n as

$$F_c^{(n)} = \frac{\tilde{F}_c^{(n)}}{\langle \tilde{F}_c^{(n)} \rangle_c} \tag{13.3}$$

$$Q_p^{(n)} = \frac{\tilde{Q}_p^{(n)}}{\langle \tilde{Q}_p^{(n)} \rangle_p} \tag{13.4}$$

and the initial conditions are $\tilde{Q}_p^{(0)} = 1\ \forall p$ and $\tilde{F}_c^{(0)} = 1\ \forall c$. The elements M_{cp} are the elements of the binary matrix M defined in the previous chapter in which the element M_{cp} is zero if the country c exports the product p, 0 otherwise.

The averages at the denominators of Eqs. 13.3 and 13.4 are performed on all the *intermediate* country fitnesses and product qualities defined by Eqs. 13.1 and 13.2 respectively. At each iteration both variables (fitness and quality) are renormalized to keep constant the total export (that is, the average export per country as the number of countries is fixed) and the average quality of products respectively. We introduce this renormalization procedure because our non linear equations have two trivial absorbing solutions in 0 and $+\infty$. By the renormalization of the space in which fitnesses and qualities lie we are able to grasp the non trivial, self-consistent and economically meaningful solution.

Firstly we note that the iterations do not change the meaning of the variables, each iteration only refines the information. As a second issue, in the computation of the quality of a product, the weight assigned to countries is inversely proportional to the fitness, in such a way we properly take into account the philosophy of capabilities. The country with small fitness dominate the sum in Eq. 13.2 as expected. We also observe that this method shares the same initial condition of the method of reflections. As a final remark this version of our method can be seen as the intensive case since the matrix M is a binary matrix which does not take into account the amount of exportation of a country. Thus the metric deriving from the binary matrix is purely non monetary, this metric intensively measures a productive system.

In addition Eqs. 13.3 and 13.4 can be rewritten in a symmetric form in the following way

$$\tilde{F}_c^{(n)} = \sum_p M_{cp} \frac{1}{B_p^{(n-1)}} \tag{13.5}$$

$$\tilde{B}_p^{(n)} = \sum_c M_{cp} \frac{1}{F_c^{(n-1)}} \tag{13.6}$$

where $B_p^{(m)} = (Q_p^{(m)})^{-1}$ can be seen as the non-quality of a product[1] anyway we prefer the notation proposed in Eqs. 13.3 and 13.4 as the interpretation of the variables are clearer.

13.1.2 Extensive Case

We can also introduce an extensive version of our method in order to also consider the export volumes. As a first step we defines in analogy with the previous case the weighted matrix whose elements W_{cp} are

$$W_{cp} = w_{cp} \tilde{M}_{cp} \tag{13.7}$$

where $\tilde{M}_{cp}$ are the elements of a matrix and they are zero if the country c does not export the product p and 1 if the country exports this product independently on the volume of the exportation and w_{cp} is defined as $w_{cp} = q_{cp}/(\sum_c q_{cp})$. We recall that q_{cp} is the raw export matrix Q which specifies the yearly amount of product p exported by the country c. According to this definition the weight w_{cp} is the fraction of export of product p owned by the country c and ranges from 0 to 1.

The extensive case reads as Eqs. 13.1–13.4 where we simply replace the matrix M_{cp} with the weighted matrix W_{cp}. Economically speaking, the intensive case gives information on the level of capabilities of a country, while the extensive case, in addition, gives information about the size of the productive systems. For instance Austria and Germany are very close in terms of the intensive metric while the fitness of Germany is about ten times the one of Austria in the extensive case even if they have reached a very similar level of capabilities. The intensive metric is somehow a *per capita* fitness while the extensive metric also measures the power and the weight of an economic system.

[1] Why B? Because the italian translation of ugliness is Bruttezza.

13.1.3 Some Remarks About Databases

The databases used in this economic study are the BACI database, edited from CEPII, and the WTF (World Trade Flows) database. The first one provides the import-export (expressed in US dollars) data from 1998 to 2007 of more than 200 countries and for over 5000 product categories (following the Harmonized System 1996 coding). Instead the second one spans from 1962 to 2000 and reports data from around 150 countries and 1200 product categories (following the SITC rev. 2 coding).

Given the non linearity of the our iterative method, the quality of the database is now crucial differently from the method of reflections. In fact an error in the database in this method would be produce an error in the final result proportional to the size of the initial error. In our non linear method instead the size of an error is amplified, especially if it occurs for poor countries for the reasons previously discussed, and can produce completely inconsistent results. Unfortunately, even if the used dataset are carefully pre-processed, it happens that country's report (especially poor countries) are not reliable. A typical problem we have encountered is due to the change of categorization of products. It happens that countries do not take care of the change and submit their annual export reports with the old categorization producing artificial drops in the export of our dataset.

Another typical problem is linked to the re-exportation of goods, say planes, in fact we find that Fiji Islands are a plane exporter while, for real, Fiji only sells used planes and does not produce them. These technical problems set some conceptual issues and a careful handling of data is required in order to obtain a consistent metric. More details about data pre-processing are given in [2].

13.1.4 Results

Let us now test our method on the toy model introduced in Sect. 12.3.2. We recall that in this toy model we construct an artificial tripartite network so that we are able to measure the number of capabilities owned by a country. In Fig. 13.1 we see that, differently from the method of reflections, the correlation between fitness and capabilities is kept constant by our method. It could be argued that the method has a better performance than the method of reflections but the iterative method does not increase the initial correlation. We recall that the tripartite network of this toy model is built starting from two random matrices therefore all the information is included in the initial step, the resulting M matrix is not triangular in this case so that there is no additional extractable information.

We have seen that the method of reflections produces a shrinkage of the information and in detail that all the variables $k_c^{(n)}$ (for even n) converges to the same values. In Fig. 13.2 we compare the two methods on the binary matrix M for the year 2000. We can see that our method, step by step, tends to enlarge the fitness distribution and the asymptotic distribution spans several order of magnitudes differently from

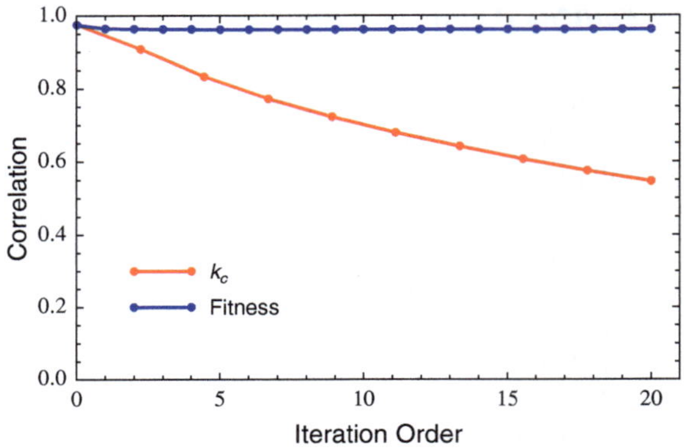

Fig. 13.1 Differently from the method of reflections the method defined by Eqs. 13.1–13.4 keeps the correlation with the endowment of capabilities constant. We argue that the iterations of the method do not increases that correlation since the artificial tripartite network is built from two random matrices therefore the projected M matrix does not have any particular structure and consequently there is no additional extractable information with respect to the zero order

the result of the method of reflections which produces a delta function as asymptotic distribution. In some sense a very broad fitness distribution is an expected feature since countries show very different degree of diversification and the complexity of a country is, roughly speaking, exponential in the number of capabilities owned by a country (see [1, 3, 4] for further details). In order to make this point clearer, let us consider USA, Nigeria. At zero order the fitness of these two countries is the diversification and we have USA's one is higher than Nigeria. According to the method of reflections the two fitnesses will be drawn up by the first iteration, instead, our first iteration will add the first order correction to the initial fitnesses, the average quality of USA's products is higher than Nigeria's one, then the distance between the two countries is enhanced. Step by step the distance between the two countries will be refined and magnified until the asymptotic value of fitness and quality is reached.

Let us now investigate the properties of the fitness and quality distribution in the intensive and extensive case. It is known that many distributions studied in Social Sciences are power law, or Pareto law as they are called in Economics [5, 6] (see also appendix for further discussions about Pareto laws). An equivalent way to investigate the properties of the distribution is the analysis of the so called rank-size plot of the phenomenon. We simply order the elements of a set with respect to their *size*, whatever is this size, and we plot the rank versus the size itself. The link between the distribution and the rank-size plot is very easy to demonstrate (see appendix). It is found that many extensive economic indicators, such as the GDP, are characterized by a rank size law named Zipf Law, that is the size $x(k)$ is inversely proportional to the rank k , i.e. $x(k) \sim 1/k$. On the other hand for intensive indicators Zipf's Law does not hold. Given this piece of information we compare the rank-size laws of the intensive and extensive country fitnesses with the GDP per capita and the

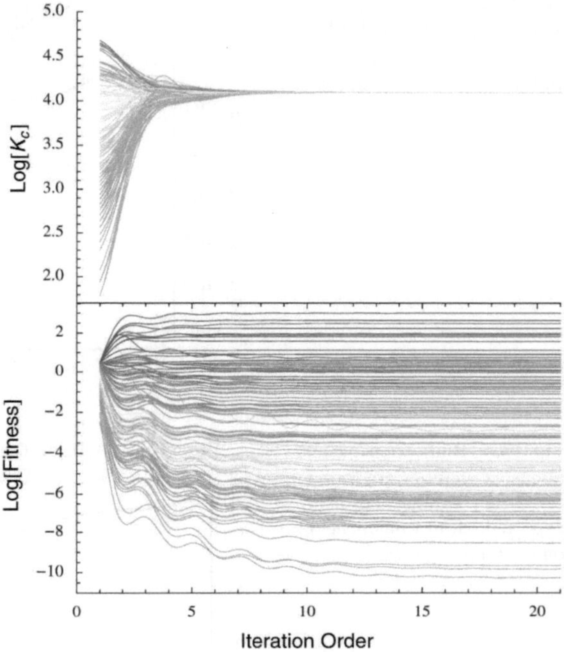

Fig. 13.2 Comparison of the method of reflections with Eqs. 13.1–13.4

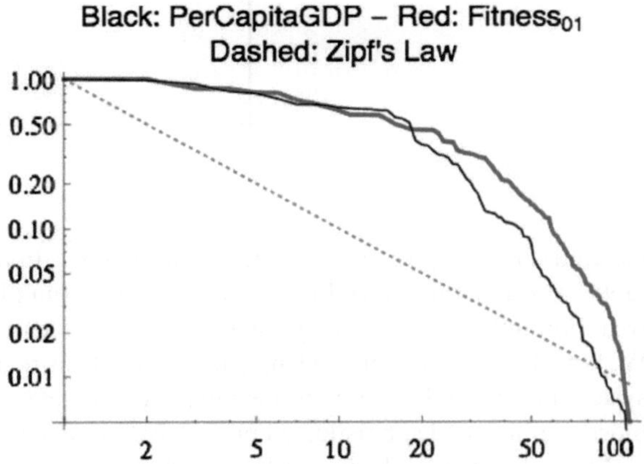

Fig. 13.3 Comparison between the GDP per capita and the intensive fitnesses

GDP respectively. As shown in Figs. 13.3 and 13.4 intensive fitnesses are in good agreement with GDP per capita and as expected they do not satisfy at all Zipf's Law. Instead extensive fitnesses matches the behavior of the GDP and for the most developed countries the rank size plot is well described by a Zipf's Law.

We observe that we do not have a traditional economic indicator to make a direct comparison with the product qualities. However we can argue that the complexity of a product, at least in the framework of capabilities, should be almost independent on

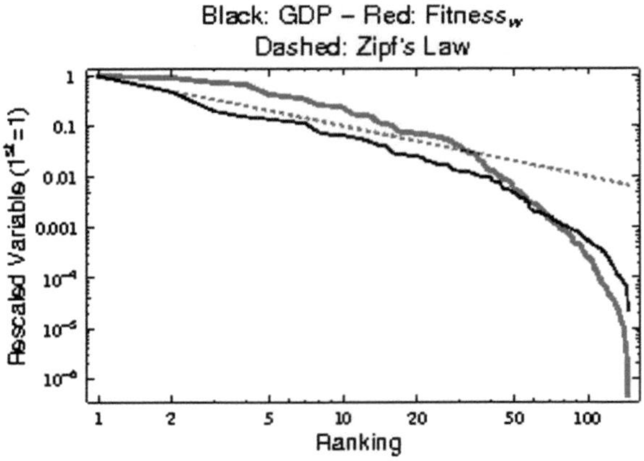

Fig. 13.4 Comparison between the GDP and the extensive fitnesses. Both are well-described by a Zipf's Law

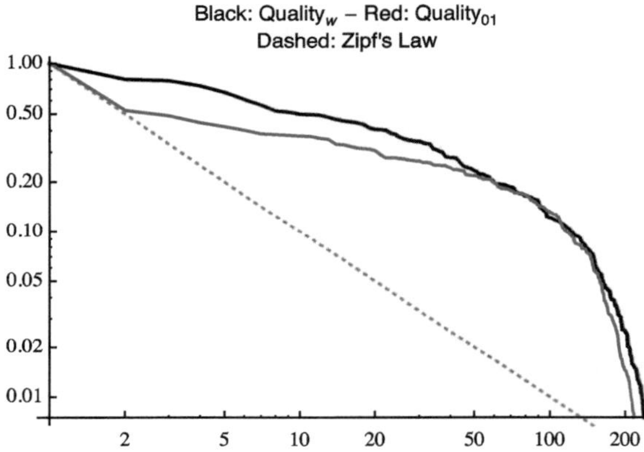

Fig. 13.5 Both extensive and intensive qualities are found to be very similar and, as expected, almost independent on the kind of method

the kind (i.e. intensive or extensive) of the analysis. In fact we find that both rank-size plots are very similar (see Fig. 13.5). As a final example of the results of our method we report in Fig. 13.6 the temporal evolution of fitnesses for some selected countries. We note the growing trend of China in both analysis revealing that China's productive system is constantly getting more complex and that Germany is the country with the most complex and sophisticated productive system according to our method.

More detailed analysis can be found in [2, 7] where, for instance, we analyze the case study represented by BRIC countries (Brazil, Russia, India and China) and we find that our metric is in agreement with traditional economic reports on this four countries but gives also non trivial insights on the potential future growth (and possible recession) of these countries.

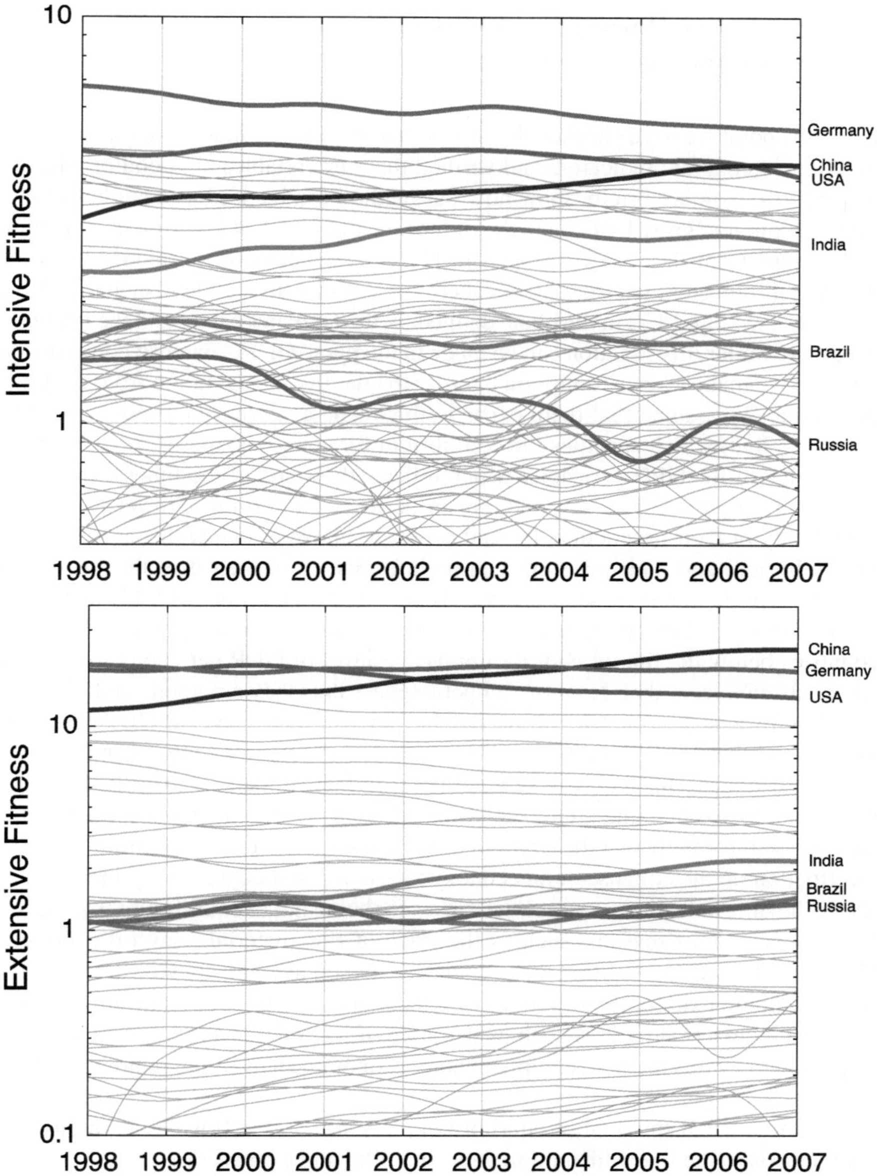

Fig. 13.6 Evolution of the fitness in the intensive and extensive case

13.2 Predictive Power: Economic Growth and Financial Applications

Until this point we have checked that fitness and quality are in good agreement with traditional economic indicators and verified that our variables are good proxies for capabilities. Now we briefly show how our method can be used as a predictive tool for the economic growth of countries and for the definition of investment strategies.

We start from the idea developed in [1], the deviation of the metric from the traditional indicators is informative. Therefore we investigate the dispersion of the countries (see Fig. 13.7) in the plane defined by the GDP per capita and the intensive fitness (a similar investigation can be made in the plane GDP—extensive fitness). At this step we make a key assumption in order to quantify the deviation from the GDP per capita. We assume that there exists in this plane a curve (our guess is that this curve is a line) which specifies the equilibrium for countries, that is there exists an ideal value of the GPD per capita which reflects the value of the fitness. Countries which stay on this curve, say line, are in perfect equilibrium with respect to the complexity of their production, instead countries above or below this curve are out of equilibrium and their positions can be informative on their growth (positive or negative). In the plane in which on y-axis we put the logarithm of GPD per capita and on the x-axis the logarithm of the fitness according our assumption we have that countries below the line of equilibrium have a level of GDP per capita lower than what expected from their level of competitiveness thus they are good candidates for growing. Instead the countries above the line have reached a level of GDP per capita higher than their *equilibrium* value, however, in that case we cannot conclude that they are candidate for a default or a collapse. In fact we believe that the dynamics below and above this line are intrinsically different. While the distances below the line can be informative on the strength of the growth of a country, a similar piece of information on the collapse of an economic system is not contained by the distance above the line. In fact we can trace the causes for being out of equilibrium from above back to exogenous reasons, such as the price of raw materials, say oil exporters. Consequently we can say that these countries live on average above what expected from the complexity of their productive system but this out of equilibrium situation is fueled by external factors, such as the price of natural gas and oil, which cannot be predicted by this method, complementary economic analysis are required in these cases.

In order to find this line of equilibrium $ax + by + c = 0$ we assume this line corresponds to the line which minimizes the following expression

$$\min_{a,b,c} \left[\sum_i f_i \left(\frac{|ax_i + by_i + c|}{\sqrt{a^2 + b^2}} \right)^2 \right] \tag{13.8}$$

where the index i = countries thus x_i is the logarithm of the fitness of the country i and y_i is the logarithm of the GDP per capita of this country and

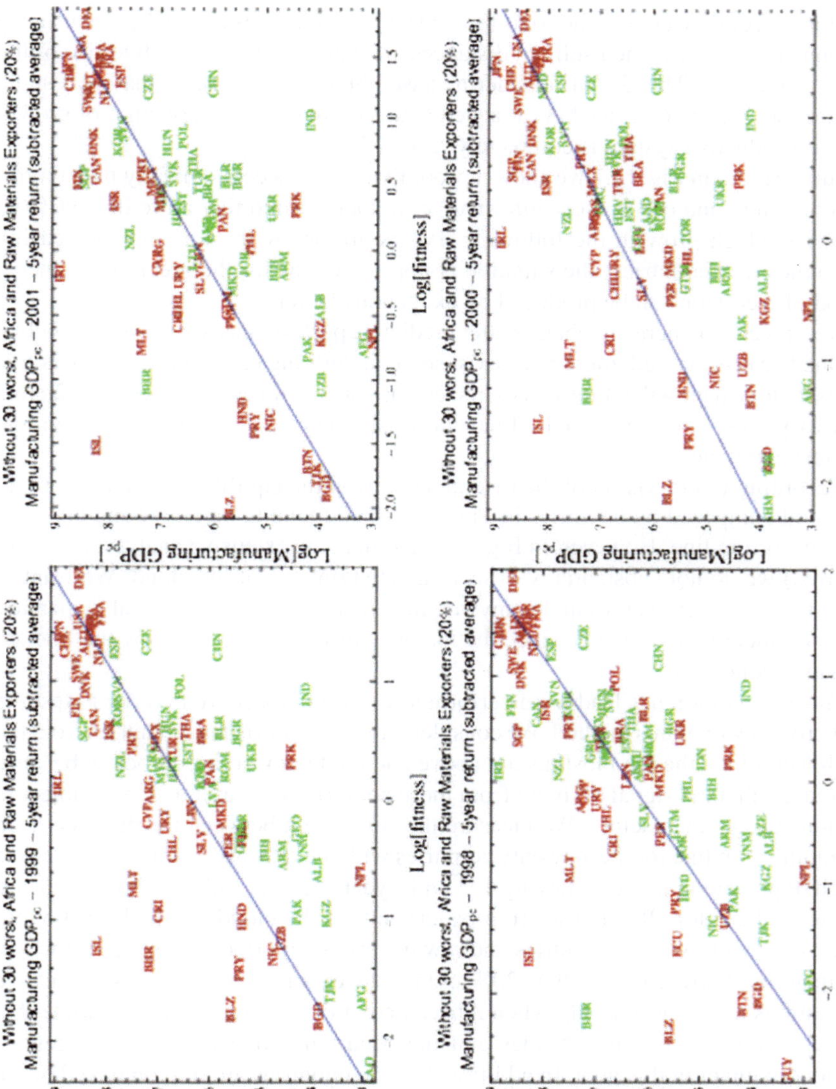

Fig. 13.7 Dispersion of countries in the plane fitness-GDP per capita. The equilibrium line appears to be very informative and predictive since more than 90 % of countries which grow in the following 3 years are below the line. For decreasing economies (*red* countries) the situation is not as clear as for the *green* ones since a period of recession can be produced by exogenous factors

$$f_i = \frac{GDP_i}{\sum_j GDP_j}. \tag{13.9}$$

The line of equilibrium we choose is the line which minimizes the euclidean distance of countries from the line itself but distances are weighted by the relative size of the GDP of countries. The idea of considering a weight for the distance is that we assume that industrialized countries have reached equilibrium since the potential of growth of their productive system is completely revealed.

Furthermore in Fig. 13.7 we show a proof that is line is economically meaningful and consistent and can be very informative. In fact we find that more than 90 % of countries which grow in the following 3 years are below the line. For decreasing economies (red countries) the situation is not as clear as for the green ones since a period of recession can be produced by exogenous factors.

We expect, in general, to obtain better predictive performances for countries below the line because, as said, there must exist some underlying economic and productive forces to drive a growth while a recession can be caused by several exogenous factors such as prices of raw materials, financial crisis, natural disasters, wars, political establishment, etc.

According to our vision of the discrepancy from the equilibrium we expect that the cloud composed by countries in Fig. 13.7 should tend to be closer and closer to the equilibrium line. However, in Fig. 13.7 and in some extensive analysis (here not reported) we do not substantially observe a net shrinkage of the cloud. We want to stress that the discrepancy can be anyway informative even if the global dispersion does not undergo a shrinkage since the majority of the countries below the line are growing economies.

This observation has lead us to perform a selection procedure in order to test the predictive power of this method. We consider only the more distant countries from the equilibrium line, the ten most distant above and the ten most distant below because we assume that the signal deriving from these countries is clearer and less subject to spurious exogenous factors. We measure the correlation between the distance from the equilibrium line for these twenty countries with the returns (i.e. the variations) of the GDP per capita in the following 1, 3 and 5 years. In Fig. 13.8 we show the result for the analysis for all countries (top panel) and for the cloud of Fig. 13.7 (bottom panel) removing African countries and raw materials exporters (countries for which raw materials represent more than 20 % of the total export). Positive and significative correlations are achieved only when African countries are removed. We argue that the very high political instabilities and the occurrence of wars make exogenous factors dominant in this area. In addition African countries are often blocked in the so-called poverty traps, that is, even if some new capabilities are added, benefits are not observed since the costs to develop these capabilities are higher than the effective economic advantages. Raw material exporters are removed because their growth (or recession) is artificially linked to the financial price of raw materials. These two aspects show that our method requires some complementary economic considerations to obtain significative results.

Higher correlations are obtained for sectorial GDP per capita (see Fig. 13.9) for which we observe correlation around 0.3–0.5 for a prediction term of 3–5 years.

In conclusion the dynamics in the plane fitness-GDP per capita appears very complicated and we plan to undertake a systematic investigation of the trajectories

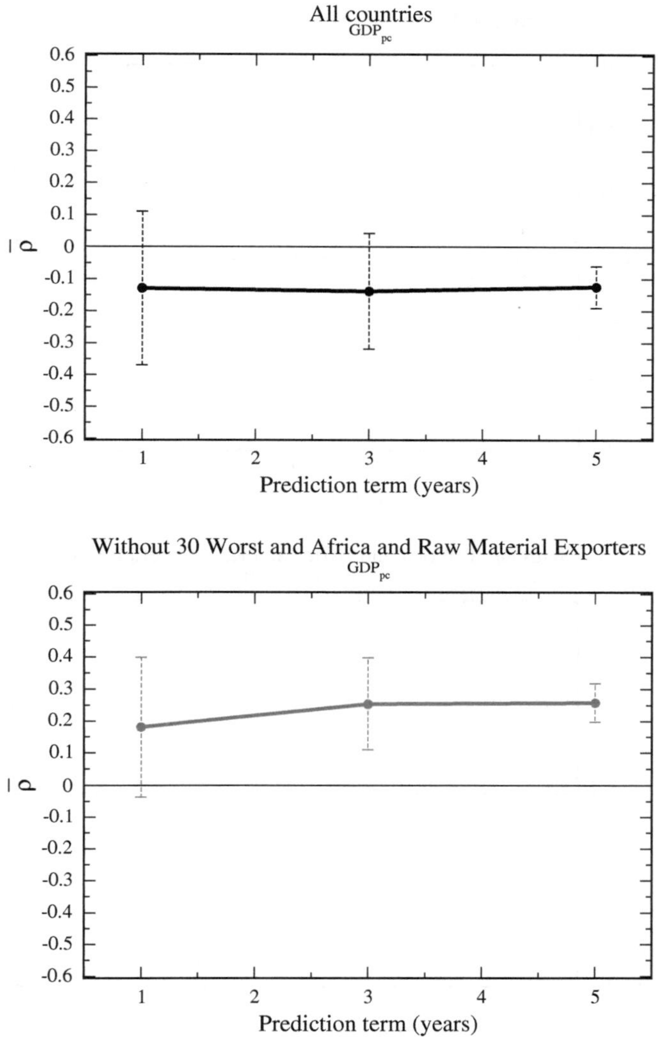

Fig. 13.8 Correlation between the distance from the equilibrium line for the most distant countries (10 above and 10 below) and the returns of the GDP per capita over the next 1, 3 and 5 years. (*Top panel*) analysis on all the countries, (*Bottom panel*) analysis performed removing African countries and countries for which raw materials represent more than 20 % of the total export

followed by countries in order to extract or formulate some evolution rules for the fitness and GDP per capita. In this way we may also individuate the poverty traps (African countries) and the richness traps (see for instance the industrialized countries in the top right part of all panels of Fig. 13.7). We can also characterize the typical path followed by emerging economic systems.

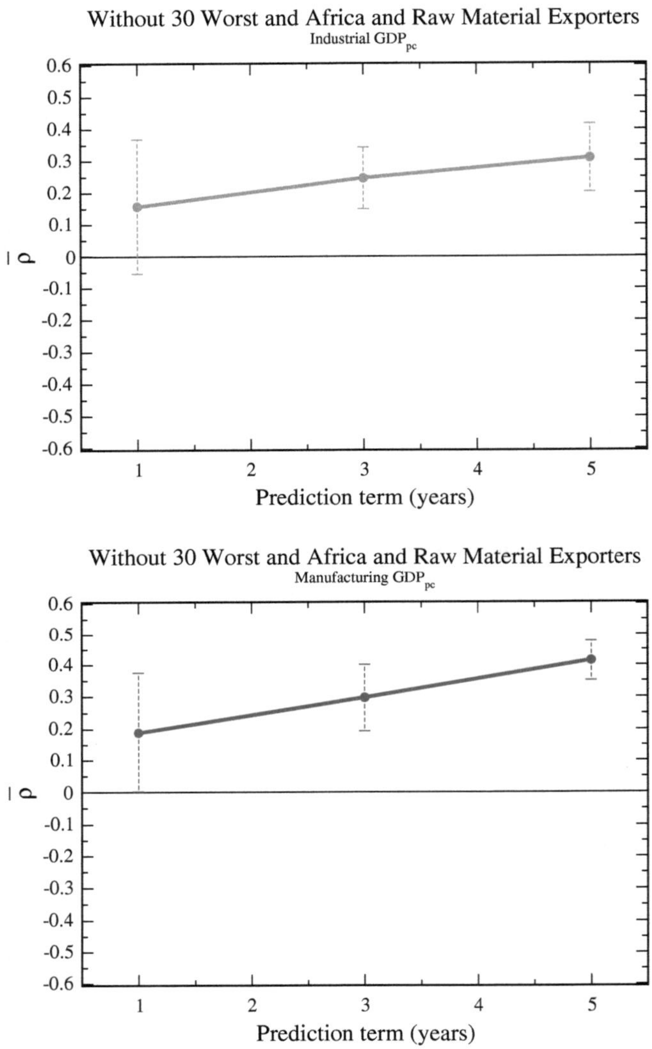

Fig. 13.9 Correlation between the distance from the equilibrium line for the most distant countries (10 above and 10 below) and the returns of a sectorial GDP per capita over the next 1, 3 and 5 years. (*Top panel*) Analysis for industrial GPD per capita, (*Bottom panel*) analysis for manufacturing GPD per capita

Financial strategies. Our method also suggests some possible financial implementations. We are currently testing some simple strategies by building a basket of stock indexes of countries below the equilibrium line and we are also testing if the distances from this line can be used to build and rebalance a weighted basket of indexes (national and sectorial).

13.3 Summary and Perspectives

In this chapter we have presented a novel method to define a self-consistent and non monetary metric for the competitiveness of the countries and the complexity of the products. We have shown that the method is in good agreement with traditional economic indicators such as the GDP per capita but, differently from the method of reflections, is also able to grasp the level of capabilities of a country.

In order to achieve this we introduce a non linear algorithm which matches the main message of the theory of capabilities, the information that a product is made by a developed country is not informative, while the information that a poor country is able to export a product means that this product requires a low level of sophistication and few capabilities. In practice in the definition of the quality of the product we weight the exporters of this product with the inverse of their fitness.

In this highly non linear framework the database becomes crucial as even minor errors in the country export records can be magnified and produces completely inconsistent results.

This method, even in its raw version, shows a good predictive power for sectorial GDP and some promising financial strategies are going to be extensively tested. In addition the iterative equations can be generalized introducing suitable exponents with respect to which the predictive power could be optimally tuned.

References

1. Hidalgo, C. A., & Hausmann, R. (2009). The building blocks of economic complexity. *Proceedings of the National Academy of Sciences*, *106*, 10570.
2. Tacchella, A., Cristelli, M., Caldarelli, G., Gabrielli, A., & Pietronero, L. (2012). A new metrics for countries' fitness and products' complexity. *Scientific Reports on Nature*, *2*, 723.
3. Hidalgo, C. A., Klinger, B., Barabási, A. L., & Hausmann, R. (2007). The product space conditions the development of nations. *Science*, *317*, 482.
4. Hidalgo, C. A., & Hausmann, R. (private communication).
5. Pareto, V. (1906). *Manual of political economy*. New York: Kelley.
6. Newman, M. E. J. (2005). Power laws, pareto distributions and zipf+s law. *Contemporary Physics*, *46*, 323.
7. Caldarelli, G., Cristelli, M., Gabrielli, A., Pietronero, L., Scala, A., & Tacchella, A. (2012). A network analysis of countries' export flows: firm grounds for the building blocks of the economy. PLoS ONE, 7, e47278. http://arxiv.org/abs/arXiv:1108.2590.

Appendix A
There Is More Than a Power Law in Zipf

The largest cities, the most frequently used words, the GDP of the richest countries, and the income of the most wealthy billionaires can be all described by a relation known as Zipf's Law. This is a power law which captures the scaling relation between the frequency or size of a very broad class of objects with respect to their rank. The implications of this rank-size rule are considerably broader than many of the cases discussed hitherto, appropriate applications requiring an additional fundamental property which we call the coherence of the sample distribution in question, corresponding to a screening between various elements of the set. A spectacular consequence is that a subset of the Zipfian set may not show any scaling characteristics whatsoever and we thus propose an analysis to extract new and useful information from this novel property [1].

Zipf's Law [2, 3], usually written as $x(k) = x_M/k$ where x is size, k is rank, and x_M is the maximum size in a set of N objects, is widely assumed to be ubiquitous for systems where objects grow in size through competition or are determined by external constraints [4–7]. These processes force the majority of objects to be small with few that are very large. Income distributions are one of the oldest exemplars first noted by Pareto [8] who considered their frequencies to be distributed as a power law. City sizes, firm sizes and word frequencies [5, 9, 10] have also been widely used to explore the relevance of such relations while more recently, interaction phenomena associated with networks (hub traffic volumes, social contacts [11, 12] also appear to mirror power law-like behavior. Zipf's law is a popular, longstanding and widely discussed phenomenon which has given rise to many more than a thousand papers [13]. It has rapidly gained iconic status as a *universal* for measuring scale and size in such systems, notwithstanding the continuing debate as to the appropriateness of the power law (or $1/k$ behavior) and the mixed empirical evidence which remains controversial [4, 5].

Here we argue that the very definition of the objects comprising the system in the first place has to be undertaken with extreme care [14]. Many real systems do not show true power law behavior because they are incomplete or inconsistent with the conditions under which one might expect power laws to emerge [15]. We show

M. Cristelli, *Complexity in Financial Markets*,
Springer Theses, DOI: 10.1007/978-3-319-00723-6,

that the origin of $1/k$ behavior is considerably more subtle than one might expect at first sight and than is usually stated in the scientific literature. In fact we report on a surprising and usually ignored property which points to the fundamental importance of the nature or the 'coherence' of the sample (or subsample) of objects or events defining systems of interest that may follow a perfect Zipf's Law or may markedly deviate from it.

A.1 Zipf's Law Works for Rich but not for "Poor": The Coherence of the Sample

Our thesis is remarkably easy to demonstrate. Consider the income of 20 people whose distribution satisfies Zipf's Law and where the maximum income $x_M = x(1)$ is \$1m. If we consider a subsample of the first 10 persons (the richest), then this subsample will certainly satisfy the same Zipf's Law. However when we consider the second group of 10 persons (the poorest), the incomes of the first two persons is \$1m/11 and \$1m/12, while the ratio of the second to the first is 11/12, very different from the first two incomes in the richest set whose ratio is 1/2. These differences apply to all the other corresponding ratios between successive objects in the two subsets. In Fig. A.1, we elaborate this example first by ranking the incomes of the 390 billionaires resident in the US in 2010 (from the Forbes List [16]) whose income, once ordered, approximately follows a Zipf's Law. This provides a highly graphic demonstration that by partitioning two sets generated from one law, two laws are necessary to explain their resulting parts. This point has extremely wide ramifications for all work on scaling systems and power laws in general, rank size and Zipfian relations in particular. There is little evidence that the importance of this point has been grasped, or if it has, it has been widely ignored. To explore its implications, from an elementary analysis of the N objects in the full sample, we select an ordered subsample of all objects below the rank $k = k^*$. We examine this set as a rank-size law where the new rank $k' = 1, 2, \ldots$ is defined in terms of the original rank k as $k' = k - k^*$. The subsample now follows the relation

$$x(k') = \frac{x'_M}{\left[(k'/k'_M) + 1\right]} \tag{A.1}$$

where the new maximum is $x'_M = x_M/k^*$. Noting that in the original set, the ratio of successive sizes is $x(k+1)/x(k) = k/(k+1)$, in the subsample this ratio is $(k^* + k')/(k^* + k' + 1)$ which shows quite clearly that the second set does not follow the same rank size rule as the initial set. In fact for the subdivision in Fig. A.1 where we divide the top 390 billionaires into the first richest 195 and the second *poorest*, we find that the ratio of the first to the second in the second set, expressed in terms of the rescaled rank $k' = k - 196$, is $196/197 \approx 0.995$ which is very different from the expected ratio for a pure Zipf's Law. In the inset, we also show the same failure for the

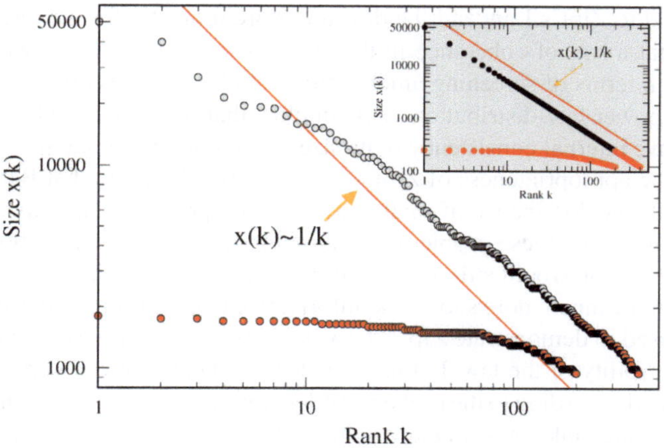

Fig. A.1 Zipf's law for the richest billionaires in the United States. The richest 390 persons in the US are billionaires whose wealth we plot against their rank as the uppermost set of points (the first 195 richest being *grey circles*, the second 195 poorest being *red circles*). The second set is the subsample that we translate to the original ranks and plot as the set of *red circle* points below the diagonal straight line which is the pure Zipf plot associated with $x(k) = xM/k$. The *inset* is a pure Zipf plot dimensioned to the entire set of 390 billionaires and the poorest subsample of 195 (*Source* Forbes List)

second ordered set (red dots) which occurs when the rank size is based on an idealized Zipf's Law but dimensioned to the same income data. An analogous problem arises if we consider two independent sets where Zipf's Law holds for each which we then aggregate. It is obvious that Zipf's Law cannot hold for the aggregated set.

Such elementary examples show in a rather dramatic way, the crucial role played by a property of internal consistency or completeness of the total set under examination which we call *coherence*. A thorough examination and some reflection on empirical applications of Zipf's Law particularly to social systems, suggests that many applications to data are based on systems where the data is incomplete in some sense [17, 18]. These issues elevate consistency in system and object definition into a new open problem which we address in this chapter. Thus for Zipf's Law to hold, a set of objects must not contain replicas of the kind just noted, nor must it be applied to a sample of objects or events that is less than the whole, unless the sampling is able to anticipate the structure of the whole. This, as we will see, is a powerful and often difficult criterion to meet.

A.2 Why We Need More Than a Power Law

When we say *There is more than a power law in Zipf*, we mean that although an underlying power law distribution is certainly necessary to reproduce the asymptotic behavior of Zipf's Law at large values of rank k, any random sampling of data does not lead to Zipf's Law and the deviations are dramatic for the largest objects. We show that a sort of coherence in the entire dataset is necessary which may be interpreted in terms of screening among different objects. This effect is beyond the underlying power law distribution. This implies that any system which obeys this law must have internal consistency in its size distribution or its sample. In our quest to explore the appropriateness of Zipf's Law, it is worth noting that Benford's Law which reveals the dominance of small numbers with properties akin to a power law, does not suffer from these problems of sampling, for any random subset, union of sets, or aggregation would still meet Benford's Law [19]. In this sense, we consider Zipf's Law to be much more subtle and informative than Benford's in that the system of interest used to demonstrate Zipf's Law is of crucial importance to the relevance, hence applicability of the law. Let us consider N objects (cities, word frequencies, etc) distributed according to the probability density $p(x) \sim x^{-\alpha}$. In sorting the size of these objects, the rank k associated with the size $x(k)$ corresponds to the probability of finding $k-1$ objects larger than $x(k)$, between $x(k)$ and the maximum value x_M. Then for rank k we can write

$$k-1=(N-1)\int_{X(k)}^{X_M} p(x)dx \approx (N-1)\frac{1}{1-\alpha}x(k)^{1-\alpha} \tag{A.2}$$

where $k=1,2,\ldots,N$ and we assume $x_M \to \infty$. From Eq. A.2, it is easy to derive the rank size law as $x(k) = C/k^{1/1-\alpha}$ from which Zipf's Law is recovered when $\alpha=2$. However, this argument only holds for large values of k because we assume $x_M \to \infty$. If we do not ignore x_M, accept that it is finite in a realistic case, and set $\alpha=2$, we then explicitly define the normalization constant C from the boundary values of the set $[x_m, x_M]$ of $p(x)$ in Eq. A.2 as $C = x_m x_M/(x_M - x_m)$. Using C, we can then define the most appropriate rank size rule for empirical analysis as

$$x(k)=\frac{C}{\dfrac{k-1}{N-1}+\frac{C}{x_M}} \tag{A.3}$$

As expected, the rank size rule in Eq. A.3 behaves asymptotically as $1/k$ but for small values of k (large values of $x(k)$) which is the region we tend to be mostly interested in, the behavior of Eq. A.3 shows a deviation from a pure Zipf's Law due to the constant term C/x_M present in the denominator. The value of this constant also sets the rank $\hat{k} = NC/x_M$ above which $x(k)$ can be approximated by $1/k$ and below which the rank size law deviates from a pure Zipf's Law.

We can make this point more cogently by underlining the fact that a mechanism which is able to recover the $1/k$ behavior only asymptotically completely misses the significant features of a Zipfian set of values. In fact the largest values of this set (i.e. those values corresponding to small values of k) are actually the main expression of what we have called *coherence* or consistency of the sample. In Fig. A.1, we have seen that the problem of sample coherence is particularly important for the biggest values with the largest value in fact defining the entire rank-size law. Therefore the rank k of an independent sampling cannot be interpreted as the breakpoint in the scale at which an adequately approximated mechanism exists to explain Zipf's Law because these values are indeed the core of the problem addressed here.

It is worth noting from Fig. A.2 that Zipf's Law works extremely well for the largest values in many phenomena as in countries where cities have developed in a more integrated manner. Nigeria is a good example which during its major growth period was relatively isolated globally (see Fig. A.2) and therefore this country exhibits a nearly perfect Zipf's Law or, with respect to our interpretation, a high degree of coherence favored by its isolated growth. For many other examples in Fig. A.2, we can also report phenomena where coherence is not expected and where indeed Zipf's Law is not observed.

In fact, many applications of Zipf's Law reveal a quite severe lack of coherence in their data and lead, as in the world city data set in Fig. A.2, to the bigger question: what is missing?

In practice in order to obtain a pure Zipf's Law, this means that the range of definition of $p(x)$ from which we perform the sampling depends on the number of elements in the sample. Thus in the framework of independent samplings, we have to set N according to the range of the pdf or the range according to N. We can see this dependence between C and N as a consequence of the coherence that a Zipfian sample must have. However, rather than adopt this somewhat artificial combination of parameters, we now argue that this coherence can be interpreted in a different context in a more fashionable and natural way.

A.3 A Simple Model for Coherence: Conditioned Sampling

Instead of varying the range of the original power law $p(x) \sim x^{-2}$, we propose that a screening or conditioning effect should be introduced into the selection procedure with respect to our framework. The basic idea behind such a concept can be exemplified using the distribution of city sizes in the US. Suppose that at a certain point, we extract *New York City* from our $1/x^2$ distribution. After such an event in a random sampling, there is still a probability that *Another New York City* could be drawn from the distribution. In reality of course, such an event cannot happen because the largest cities screen one another with respect to their growth dynamics. A simple way to introduce such an effect is to make the sampling conditional. Then after a certain value is extracted, a section of the distribution around this value is thence excluded from the density. We show this schematically in Fig. A.3. In essence, we draw the size of the first object x_1 from the density $p(x) \propto 1/x^2$ that is normalized over the range $[x_m, x_M]$. The section to remove around x_1 varies from $x_{1,min}$ to $x_{1,max}$ and these bounds are computed so that the area of the removed section is A

$$A/2 = C \int_{x_1}^{x_{1,max}} p(x)dx \Rightarrow x_{1,max} = 2x_1C/(2C - Ax_1) \qquad \text{(A.4)}$$

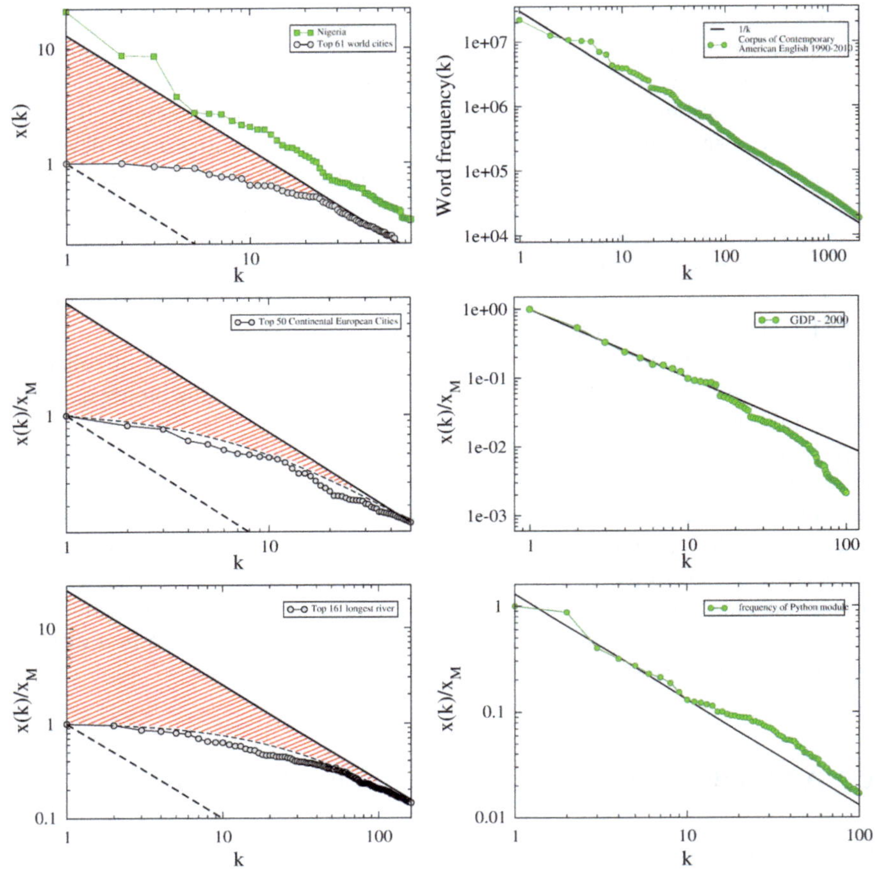

Fig. A.2 Rank-size laws illustrating different degrees of coherence; *Top Left* Rank-size for the largest world cities showing the absence of truly global cities which have developed in relation to all other cities. Therefore the world is not a coherent/fully integrated system and represents the wrong scale at which city size samples must be aggregated to obtain a Zipf's Law. Instead Nigeria (*green solid line*) shows a nearly perfect Zipfian behavior. Nigeria which is separate from the rest of Africa, represents a city system which has developed more uniformly in a more integrated fashion, The Nigeria rank size law has been rescaled for clarity; *Centre Left* Rank-size of the fifty largest European continental cities (i.e. the European part of Russia and UK are excluded). As in the case of the world cities, we observe absence of coherence at this geographical scale; *Bottom Left* River formations are mainly due to geographical and morphological constraints on the Earth's surface. Hence a Zipf's Law is not expected and in fact the river rank size rule is well approximated by the curve (*dashed black line*), predicted by an independent sampling procedure without any screening effect; 10; *Top Right* For the frequency of words in the Corpus of Contemporary American English, a quasi-perfect Zipf's Law is observed over the 2000 (and more) most used words. Linguistic systems are fully coherent with respect to our interpretation of Zipf's Law. *Centre Right* If we rank the Gross Domestic Product (GDP) of world countries, we observe a Zipfian behavior for the 30 richest. *Bottom Right* As for words, a Zipf's Law also appears in the frequency of usage of Python modules in a computer science project domain

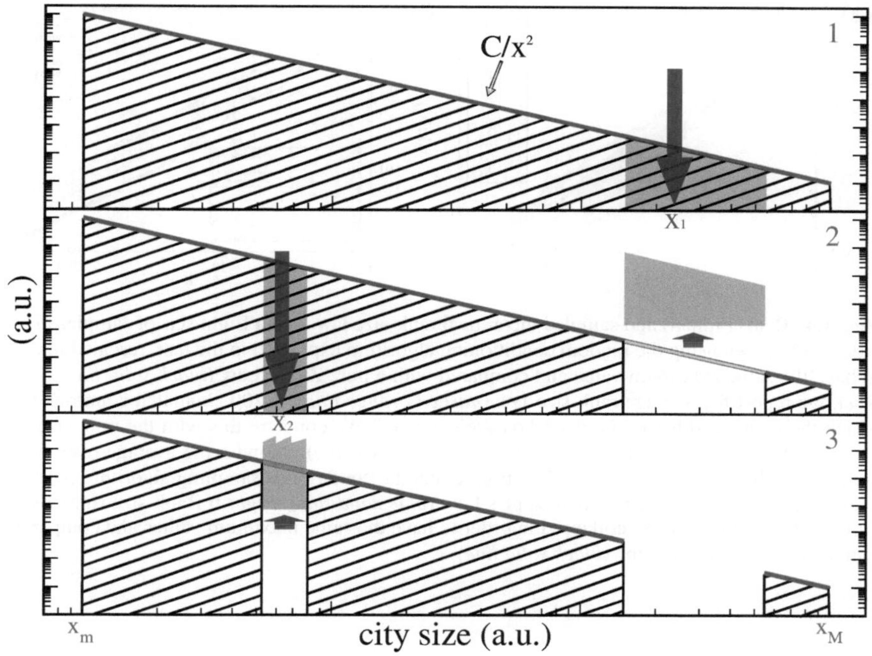

Fig. A.3 Successive conditioned draws. When we extract an object (or a city size) such as x_1, we remove a section (*blue slice* in panel *1*) of the probability density around the drawn value. See the text for the details of how to compute the slice to remove. Then the density moves to the reduced distribution in panel *2* from which the next object is drawn in the same way with the slice associated with x_2 being removed in panel *3*

$$A/2 = C \int_{x_{1,min}}^{x_1} p(x)dx \Rightarrow x_{1,max} = 2x_1 C/(2C + Ax_1) \tag{A.5}$$

The area A of the forbidden section is a priori arbitrary and we fix it to be equal to $1/N$ where N is the total number of extractions. This slice $[x_{1,min}, x_{1,max}]$ is then removed meaning that the subsequent object of size x_2 must be not be drawn from this area. The computation proceeds recursively in this fashion until the required number of objects has been sampled as implied in Fig. A.3.

In Fig. A.4a, we show a series of samples, normalized with respect to their maximum values where the scaling is close to Zipf's Law but where their position, hence actual populations are heavily influenced by the lower ranked, larger sized draws. In Fig. A.4b, we show real data which corresponds to the city size distributions for several different countries. The sampled and real distributions in Fig. A.4 are sufficiently different en masse to indicate that many real city size distributions are incoherent in comparison to the theoretical equivalents. In Fig. A.4b, there are some countries such as the UK, Russia, Iran and to a lesser degree France where the capital cities exercise

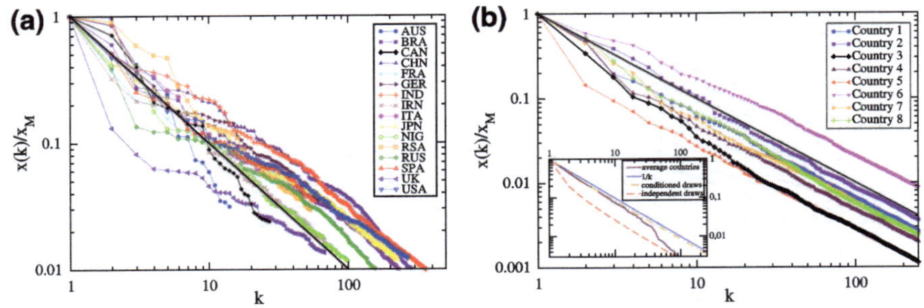

Fig. A.4 Real (Zipfian) and sampled theoretical rank-size law. *Left* **a** Eight sets of samples of 250 cities each drawn using the random conditioning algorithm explained in the text, rescaled in order to have the same maximum value and compared with a pure Zipf's Law (*black solid line*). In the *inset*, we report the average rank-size law (*dashed orange line*) of 200 simulated countries from which the eight reported in the main box are extracted. We compare this with the rank-size rule produced by independent random samplings (*red dashed line*) and with the average over the 16 countries of **b** (*maroon solid line*). We observe that the conditioned sampling algorithm produces a striking result very close to a pure Zipf's Law (*blue solid line*). *Right* **b** Rank-size rules for the cities in 16 world countries collapsed in order to have the same maximum values and comparable with Zipf's Law for the same (*black solid line*)

a primate city effect which indicates extreme concentration compared to other elements in their size distributions. Explanations for these deviations are loose: cities serving empires beyond their national boundaries, and highly centralized administrations, are obvious explanations. Most other countries reveal the opposite in that their largest cities have lesser sizes than might be expected if Zipf's Law were to play out exactly. We also consider that screening of one object with respect to another occurs at different hierarchical levels. Thus we consider that conditional sampling of the data and exploration of the extent to which cities screen one another is key to an understanding of city size relations.

A.4 Summary and Perspectives

This situation forces conceptual problems of a new type because up to now, most researchers dealing with this problem have attempted to develop a theory for Zipf's Law which is to be found in the underlying distribution $1/x^2$. In fact we now see clearly that such a theory cannot be developed without considering the problem of the sample coherence which in cities, income distributions and in many systems that have been considered as being described by power laws, will always show itself up as the phenomenon we have referred to as screening. The question of defining each individual object also effects the coherence of the system because if objects are split, disaggregated, or indeed merged, aggregated, their order changes. As we consider

Zipf's Law to be the ultimate signature of an integrated system, it is thus important to devise general models which include coherence in a simple but generic way. In this line of reasoning, coherence and screening could be the result of some kind of optimization in growth processes or of an optimal self-organization mechanism of the system with respect to some (finite) resources. This must be the next step in providing further new and novel perspectives on this entire area of study.

References

1. Cristelli, M., Batty, M., & Pietronero, L. (2012). There is more than a power law in Zipf. *Scientific Reports on Nature, 2*, 812.
2. Zipf, G. K. (1949). *Human behavior and the principle of least effort*. Cambridge, MA: Addison-Wesley.
3. Saichev, A., Malevergne, Y., & Sornette, D. (2010). *Lecture notes in economics and mathematical systems* (Vol. 632). Heidelberg, Germany: Springer.
4. Newman, M. E. J. (2005). Power laws, Pareto distributions and Zipf's law. *Contemporary Physics, 46*, 323.
5. Clauset, A., Shalizi, C. R., & Newman, M. E. J. (2009). Power-law distributions in empirical data. *SIAM Review, 51*, 661.
6. Blank, A., & Solomon, S. (2000). Power laws in cities population, financial markets and internet sites: Scaling and systems with a variable number of components. *Physica A, 287*, 279.
7. Lotka, A. J. (1926). The frequency distribution of scientific production. *Journal of the Washington Academy of Sciences, 16*, 317.
8. Pareto, V. (1906). *Manual of political economy*. London: Macmillan.
9. Gabaix, X. (1999). Zipf's law for cities: An explanation. *Quarterly Journal of Economics, 114*, 739.
10. Axtell, R. L. (2001). Zipf distribution of U.S. firm sizes. *Science, 293*, 181.
11. Barabasi, A.-L., & Albert, R. (1999). Emergence of scaling in random networks. *Science, 286*, 509.
12. Barabasi, A.-L. (2005). The origin of bursts and heavy tails in human dynamics. *Nature, 435*, 207.
13. Google scholar returns just over 1000 papers with Zipf in the title and over 40,000 with reference to Zipf. Accessed 11 April 2011.
14. Perline, R. (2005). Strong, weak and false inverse power laws. *Statistical Science, 20*, 68.
15. Montroll, E. W., & Schlesinger, M. F. (1982). On 1/f noise and other distributions with long tails. *Proceedings of the National Academy of Sciences of the United States of America, 79*, 3380.
16. Forbes lists. (2010).
17. Eeckhout, J. (2005). Gibrat's law for (all) cities. *American Economic Review, 94*, 1429.
18. Rose, A. K. (2006). Cities and countries. *Journal of Money, Credit and Banking, 38*, 2225.
19. Pietronero, L., Tosatti, E., Tosatti, V., & Vespignani, A. (2001). Explaining the uneven distribution of numbers in nature: The laws of Benford and Zipf. *Physica A, 293*, 297.

About the Author

Matthieu Cristelli is a post-doc researcher at Institute for Complex Systems (ISC-CNR) in Rome from January 2012. He graduated *cum laude* in Physics in 2008 and got his doctorate in Physics at Sapienza University of Rome in 2011. He mainly works on interdisciplinary application of Statistical Physics and Complex Systems to Economic and Financial Systems and Social Sciences. In detail the main highlights of his research are Economic Complexity, Order Book dynamics (modeling and data analysis), agent-based modeling for Financial Markets, Self-Organization of markets, web data analysis, etc. He published about 20 papers on international peer-reviewed journals and he has been invited to some international conferences and participated to several ones as contributed speaker. He is involved in several projects such as EU FET OPEN project FOC Forecasting Financial Crisis and the italian PNR CRISIS LAB.

M. Cristelli, *Complexity in Financial Markets*,
Springer Theses, DOI: 10.1007/978-3-319-00723-6,

He also won a prize for his Master's Degree and his PhD thesis has been awarded and published by Springer.

In his spare time, Matthieu is an amateur photographer, a *professional* taster of foreign cuisines and a travel addicted!

Further details on his research activity can be found at http://sites.google.com/site/matthieucristelli/ while for a glimpse on his spare time have a look at http://www.thecuttingedgegallery.com.

Matthieu Cristelli

Complexity in Financial Markets

Modeling Psychological Behavior in Agent-Based Models and Order Book Models

Doctoral Thesis accepted by the Sapienza University of Rome, Italy

Springer

Author
Dr. Matthieu Cristelli
Department of Physics
ISC-CNR
Istituto dei Sistemi Complessi
Sapienza University of Rome
Rome
Italy

Supervisor
Prof. Luciano Pietronero
ISC-CNR
Istituto dei Sistemi Complessi
Rome
Italy

ISSN 2190-5053 ISSN 2190-5061 (electronic)
ISBN 978-3-319-00722-9 ISBN 978-3-319-00723-6 (eBook)
DOI 10.1007/978-3-319-00723-6
Springer Cham Heidelberg New York Dordrecht London

Library of Congress Control Number: 2013939572

Printed on acid-free paper

Springer is part of Springer Science+Business Media (www.springer.com)

MIX
Papier aus verantwortungsvollen Quellen
Paper from responsible sources
FSC® C105338

If you have any concerns about our products,
you can contact us on
ProductSafety@springernature.com

In case Publisher is established outside the EU,
the EU authorized representative is:
Springer Nature Customer Service Center GmbH
Europaplatz 3, 69115 Heidelberg, Germany

Printed by Libri Plureos GmbH
in Hamburg, Germany